T0296223

Technologien für die intelligente Automation

Technologies for Intelligent Automation

Band 2

Weitere Bände in dieser Reihe
http://www.springer.com/series/13886

Ziel der Buchreihe ist die Publikation neuer Ansätze in der Automation auf wissenschaftlichem Niveau, Themen, die heute und in Zukunft entscheidend sind, für die deutsche und internationale Industrie und Forschung. Initiativen wie Industrie 4.0, Industrial Internet oder Cyber-physical Systems machen dies deutlich.
Die Anwendbarkeit und der industrielle Nutzen als durchgehendes Leitmotiv der Veröffentlichungen stehen dabei im Vordergrund. Durch diese Verankerung in der Praxis wird sowohl die Verständlichkeit als auch die Relevanz der Beiträge für die Industrie und für die angewandte Forschung gesichert.
Diese Buchreihe möchte Lesern eine Orientierung für die neuen Technologien und deren Anwendungen geben und so zur erfolgreichen Umsetzung der Initiativen beitragen.

Herausgegeben von
inIT - Institut für industrielle Informationstechnik
Hochschule Ostwestfalen-Lippe
Lemgo, Germany

Henning Trsek

Isochronous Wireless Network for Real-time Communication in Industrial Automation

Henning Trsek
inIT - Institut für industrielle Informationstechnik
Hochschule Ostwestfalen-Lippe
Lemgo, Germany

Technologien für die intelligente Automation
ISBN 978-3-662-49157-7 ISBN 978-3-662-49158-4 (eBook)
DOI 10.1007/978-3-662-49158-4

Library of Congress Control Number: 2016930557

Springer Vieweg

Printed on acid-free paper

Springer Vieweg is a brand of Springer Berlin Heidelberg
Springer Berlin Heidelberg GmbH is part of Springer Science+Business Media
(www.springer.com)

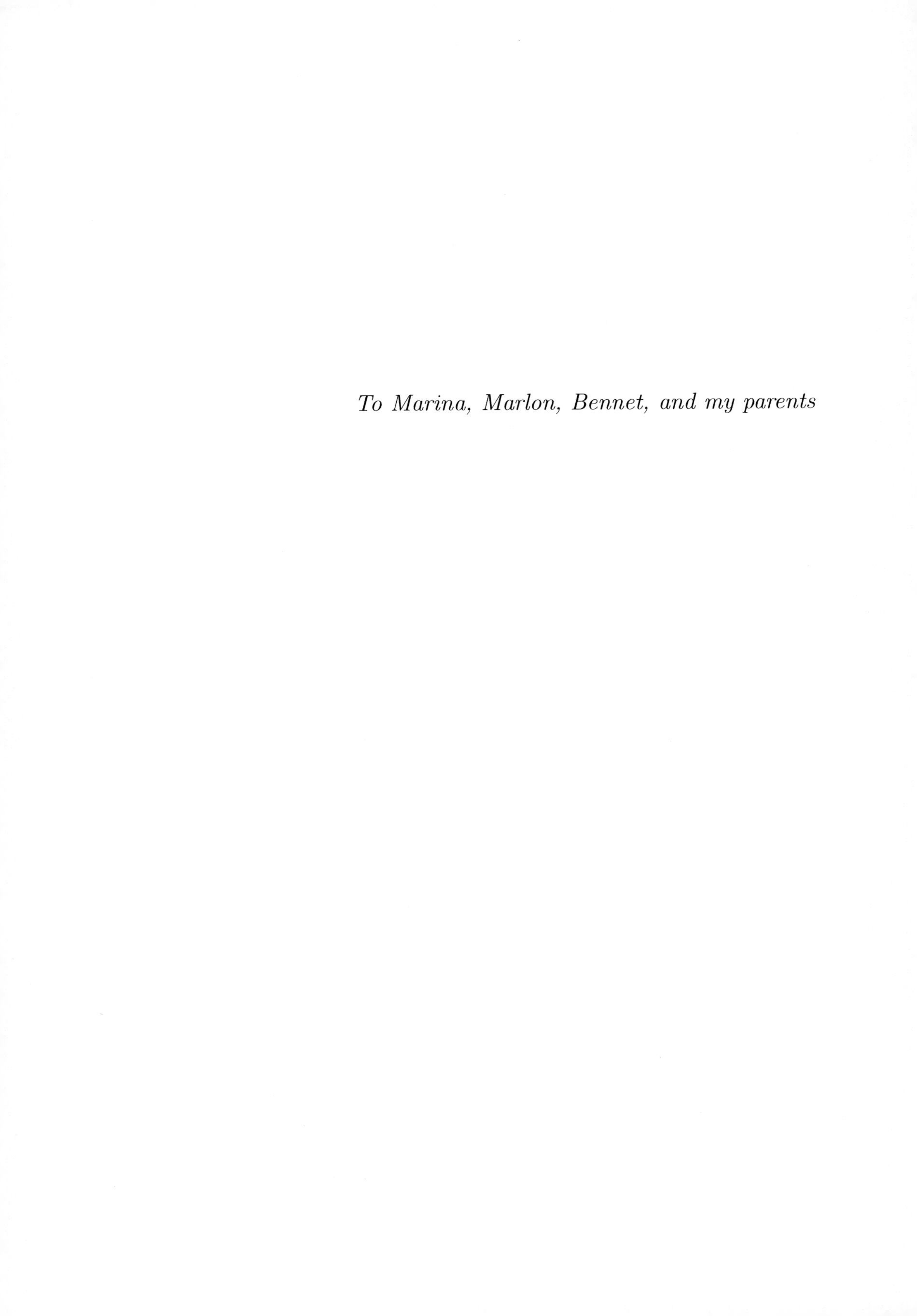

To Marina, Marlon, Bennet, and my parents

Preface

Zusammenfassung

In der industriellen Automation werden zunehmend drahtlose Technologien eingesetzt, um Anwendungen zu ermöglichen, die bewegliche Komponenten wie rotierende Anlagenteile beinhalten oder eine gesteigerte Flexibilität erfordern. Eine Vielzahl von Anwendungen, beispielsweise drahtlos vernetzte regelungstechnische Systeme, können jedoch aufgrund ihrer hohen Anforderungen an die Echtzeitfähigkeit der Datenkommunikation noch nicht oder nur mit Einschränkungen umgesetzt werden.

Das Ziel dieser Arbeit ist daher die Realisierung eines isochronen drahtlosen Kommunikationssystems für echtzeit-kritische regelungstechnische Anwendungen der industriellen Automation. Die wesentlichen Herausforderungen resultieren aus den eingesetzten Medienzugriffsverfahren, dem gemeinsam genutzten Medium und dessen begrenzter Kapazität sowie dem asynchronen Verhalten von drahtlosen und drahtgebundenen Systemen, welches das Zeitverhalten solcher hybrider Netzwerke erheblich verschlechtert.

Basierend auf den ermittelten Anforderungen von realen industriellen Anwendungen und der Charakterisierung des industriellen Funkkanals für das Anwendungsszenario, wird ein Lösungsansatz vorgestellt, der einen koordinierten, TDMA-basierten Medienzugriff verwendet, eine dynamische Ressourcenzuweisung erlaubt und die Etablierung einer globalen Zeitbasis im drahtlosen und drahtgebundenen Netz ermöglicht. Die globale Zeitbasis erlaubt eine synchrone Integration des drahtlosen Systems in bestehende Echtzeit-Ethernet-Netzwerke.

Eine prototypische Implementierung der Lösung sowie eine Simulationsstudie werden zur Evaluierung des Lösungsansatzes in zwei Fallstudien herangezogen. Die prototypische Implementierung wird für die Evaluierung in einer realen Umgebung der Fertigungsautomatisierung und für die Validierung des Simulationsmodells eingesetzt. Aufgrund der begrenzten Skalierbarkeit der prototypischen Implementierung, wird eine weitere Fallstudie anhand einer realistischen Simulationsstudie durchgeführt. Das Simulationsmodell zeichnet sich unter anderem durch eine reale Modellierung des drahtlosen Kanals aus, die auf den Ergebnissen der durchgeführten Kanalcharakterisierung basiert.

Die vorliegenden Evaluierungsergebnisse zeigen, dass mit dem vorgestellten Lösungsansatz Latenzzeiten im Bereich $\leq 10\,\mathrm{ms}$ mit einem maximalen Jitter $\leq 100\,\mu\mathrm{s}$ möglich sind, wenn alle Komponenten aktiv sind. Der Lösungsansatz erlaubt daher den Einsatz in regelungstechnischen Anwendungen, die den genannten Anforderungen entsprechen. Sobald einzelne Komponenten deaktiviert werden, wird die Leistungsfähigkeit erheblich reduziert und die Anforderungen können nicht mehr erfüllt werden.

Abstract

In industrial automation systems the deployment of wireless technologies is more and more common. This is mainly due to applications which consist either of moving components, for example rotating machine parts, or require a high degree of flexibility. However, due to their high real-time requirements the implementation of applications, such as wireless networked control systems (NCS), is rather limited or even impossible with existing wireless technologies.

The objective of this dissertation is to design an isochronous wireless network for industrial control applications with guaranteed latencies and jitter. The main challenges are a non-deterministic medium access of existing systems, the uncontrolled, shared wireless medium and its limited capacity, as well as the asynchronous behaviour of wireless and wired communication systems degrading the temporal behaviour of such hybrid systems.

Based on the requirements analysis of real industrial applications and the characterisation of the wireless channel for the application scenario, a solution approach is presented consisting of a deterministic, TDMA-based medium access control, a dynamic resource allocation and the provision of a global time base for the wired and the wireless network. The global time base allows a seamless and synchronous integration into existing wired Real-time Ethernet systems.

An implementation prototype of the proposed wireless system and a simulation case study are used for the evaluation of the solution approach in two case studies. The prototype is used for the evaluation in a real factory environment and for the validation of the simulation model. Due to given scalability constraints of the prototype, a second case study based on a realistic simulation model is conducted. A realistic channel model for the simulation, implemented based on the channel characterisation, allows more realistic simulation results.

The obtained evaluation results show that latencies $\leq 10\,\text{ms}$ and a maximum jitter $\leq 100\,\mu\text{s}$ can be achieved with the presented solution approach as long as all components are active. Thus the solution approach allows a deployment within NCSs with the given requirements. As soon as components are deactivated, the behaviour of the network is significantly degraded and the requirements cannot be satisfied any more.

Acknowledgement

First of all, I would like to thank Prof. Edgar Nett for his continuous support, many fruitful and enlightening discussions and his critical feedback to my work. I am very thankful for his guidance during the whole duration of this thesis.

The inspiring discussions and many hints as well as the continuous support from Prof. Jürgen Jasperneite throughout this work are greatly appreciated. I would like to express my gratitude to him for providing an outstanding working environment.

I am also thankful to Prof. Wolfgang Kastner, who kindly agreed to serve as third reviewer and provided several useful hints and comments.

I would like to express my gratitude to my colleagues of the Institute Industrial IT in Lemgo for many helpful discussions about details of this work and the friendly working environment. I am especially thankful to my colleagues Stefan Schwalowsky, Tim Tack, Lukasz Wisniewski, Uwe Mönks, Björn Czybik, Dimitri Block, Sebastian Schriegel, Markus Schumacher, Roman Just, Olaf Graeser, Omid Givehchi, Lars Dürkop, Jan Deppe, and Ivan Dominguez for many inspiring ideas and for superb discussions and support on technical details. I am also grateful to Prof. Uwe Meier for sharing his expertise in wireless communication in several discussions. Furthermore, his support and mentoring in my early days as a student paved the way to successfully finish this thesis. Further, I am thankful to Prof. Volker Lohweg for his helpful and friendly advices.

The research seminar discussions at the University of Magdeburg were always inspiring and an important forum for me to exchange and discuss my ideas with other outstanding experts in the wireless research area. Many thanks to Timo Lindhorst, and Georg Lukas, as well as to all other members who frequently participated in the research seminar and provided their helpful feedback.

Moreover, a big thanks goes to my colleagues of the Center for Integrated Sensor Systems, especially Georg Gaderer, Aneeq Mahmood, Patrick Loschmidt, Felix Ring, and Reinhard Exel for their great support in all questions related to clock synchronization. The wireless group of Phoenix Contact, especially Andreas Pape, provided the opportunity for wireless channel measurements on a real wind energy plant as well as the possibility to analyse a real manufacturing system in operation. Their support was very important and is greatly appreciated.

I would also like to thank my family Marina, Marlon, and Bennet as well as my parents for their patience and enduring outstanding support, especially during the last weeks, days and endless nights to finish this work. Whenever I was in doubts, they were my main source of inspiration and always a driving force.

In the final phase of this work, I was a PhD scholarship holder of the University of Applied Sciences Ostwestfalen-Lippe, and I am thankful for this helpful financial support. I am also thankful for the support of my employer, rt-solutions.de GmbH, during the final phase of this thesis.

Lemgo, June 2015 *Henning Trsek*

Contents

List of Abbreviations

1 Introduction

Wireless communication systems are almost ubiquitous and can be found in many application areas today. The advantages of using wireless solutions in all sorts of applications including residential, office, and industrial are considerable and include: major flexibility, reduced installation and maintenance costs, and increased mobility. Due to these advantages as compared to wired networks, a great tendency towards deploying wireless networks can also be observed in industrial automation systems. Especially, automation applications with either mobile entities or with moving parts can greatly benefit from wireless communication [131]. Moreover, completely new and innovative applications, which are even impossible without wireless solutions, might be realized. However, it has to be taken into consideration that industrial applications have specific requirements regarding real-time communication and reliability, which must be implicitly met for a successful deployment.

In this dissertation an isochronous wireless communication system is proposed, addressing these challenging requirements of industrial applications. In the first chapter, the targeted application scenario for this dissertation is described by two different examples. The scenario is used to define requirements and to identify the existing problems for wireless solutions in industrial automation. Finally, the solution approach addressing the derived problems is briefly introduced.

1.1 Application Scenario

Existing industrial automation systems are usually categorized according to their deployment considering the structure of the automation pyramid [171] as shown in Fig. 1.1. The automation pyramid integrates all levels of an enterprise, i. e., from the business processes down to the factory floor, into a hierarchical architecture to partition and to structure the information processing in order to cope with the given complexity [62].

The automation pyramid abstracts the whole enterprise with respect to three levels and is used in different fields of automation applications. The three levels can be separated by their different functionalities as well as their communication characteristics (transmission frequency, amount of data, real-time requirements). The transmission frequency is very high at the lowest level and decreases towards the top of the pyramid, whereas the amount of data to be transferred is at its maximum at the highest level and decreases towards the bottom. While the real-time behaviour is not important at the top level, it is of vital importance at the lowest level.

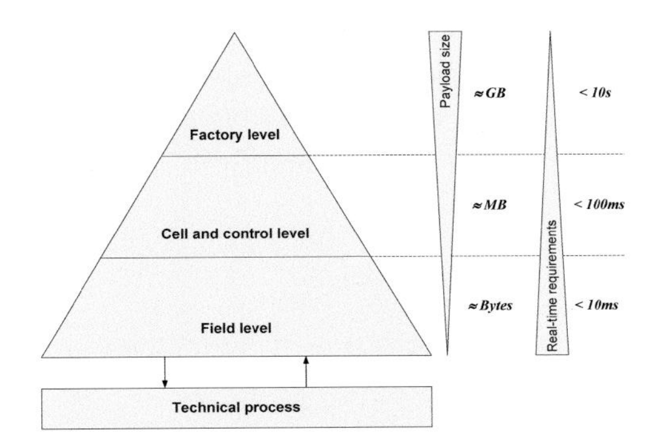

Fig. 1.1: The automation pyramid

The *factory level* is dominated by Enterprise Resource Planning (ERP) systems. They are in charge of controlling the overall business processes and the supply chain management. On this level, process data is only required for accounting, production planning, etc., which is provided by the systems of the underlying level. Usually, standard IT infrastructure is used for communication networks, since no data with real-time constraints has to be transferred. The ERP system works only with a coarse degree of detailing and cannot influence the real manufacturing process.

Therefore, the Manufacturing Execution System (MES) at the *cell and control level* are managing the manufacturing process itself and provide an important functional addition. They are directly planning and controlling the manufacturing process and ensure the process transparency [173]. Control functions also reside on the lower part of this level. They are executed by Programable Logic Controllers (PLCs) and directly influence the field level, resulting in higher real-time requirements as compared to the factory level.

The *field level* is the lowest level of the automation pyramid and provides the physical interface to the actual technical process by means of sensors and actuators. It has the highest temporal requirements for any deployed communication network, because of the direct interaction with the manufacturing process. On this level, a very small amount of payload data has to be transferred, because digital sensor values or commands to actuators consist of a few bytes only. System components at the field level, i. e., decentralized peripherals, sensors, and actuators, mainly exchange cyclic traffic with PLCs, because the control loops used within the industrial process require a cyclic behaviour. Additionally, the reliability of data transmissions must be considered, since communication errors on the field level might cause a severe damage to the plant or even endanger the life of humans.

On all levels three different types of functions must be supported by a communication network, which are control, diagnostics, and safety [126]. Whereas the third type, safety, is not part of this work, the other two types are considered with a focus on the more challenging part of control.

Control networks are deployed on the field and the control level to connect the controller, usually a PLC, to its decentralized peripherals, sensors and actuators in order to form a closed-loop control system. Together with the system to be controlled, i. e., the technical process, it is being called an Networked Control System (NCS). Its generic architecture is shown in Fig. 1.2.

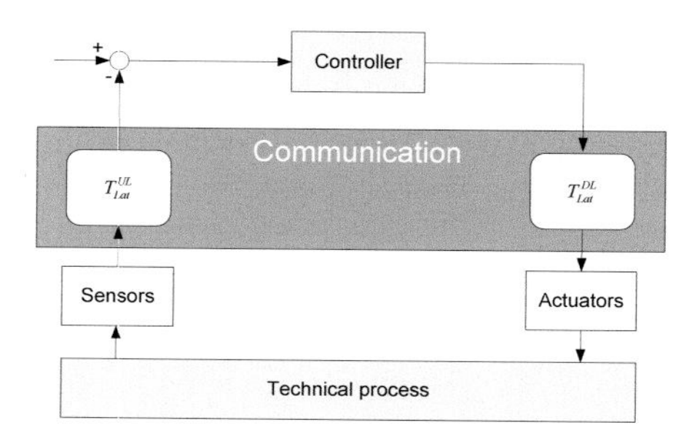

Fig. 1.2: Generic architecture of a Networked Control System (NCS)

The main advantage of NCSs is the possibility for a spatial distribution and their flexibility. Therefore, they are considered as the motivating application scenario for this dissertation. NCSs are amongst the most challenging industrial applications, because the stability of the control loop highly depends on the time varying behaviour of the network. This is illustrated as uplink latency T_{Lat}^{UL} for the link between sensors and the PLC and as downlink latency T_{Lat}^{DL} for the link between the PLC and the actuators. The overall quality of the control loop turns out to be severely affected by the properties of the network [50]. Unreliability, due to effects as packet loss or jitter, has a negative influence and is critical in this context. Several research activities in this area are investigating new approaches (cf. [116, 103, 23]) for dealing with the time varying behaviour of communication networks. However, a network still has to provide real-time guarantees and sufficient reliability when used in an NCS, both depending on the specific application. Existing generic real-time classes for control networks are shown in Table 1.1 listing their specific temporal requirements as well as the type of control as defined in [37]. Real-time class 1 (cf. Table 1.1) is used for the communication between PLCs and Human-Machine Interfaces (HMIs), i. e., whenever humans are involved in the control loop. Real-time class 2 (cf. Table 1.1) is required for communication between PLCs and decentralized peripherals. The highest requirements are imposed by real-time class 3 (cf. Table 1.1), because it addresses motion control applications to synchronize axes.

Table 1.1: Real-time classes and their requirements [80, 37]

Type of control	**Level**	**Real-time requirements**		**Real-time**
		Latency	**Jitter**	**class**
Monitoring and human control (PLC to PLC)	Control/Field	10 - 100 ms	–	1
Process control (PLC to decentralized peripherals)	Control/Field	1 - 10 ms	$\leq$ 1 ms	2
Motion control	Field	< 1 ms	$\leq$ 1 µs	3

In general, wireless solutions are interesting for networked control loops in all applications where the communication link must be established between moving or mobile system components. The type of mobility can be described as (i) rotating components, such as the following wind energy example, (ii) linear movements, like in automated storage and retrieval systems, and (iii) mobile components following a given path, e. g., Automatic Guided Vehicles (AGVs) as used in discrete manufacturing for logistics.

In order to identify existing problems for industrial wireless solutions and to derive valid requirements, the application scenario of wireless NCSs is further divided into two different application categories. The first category is characterised by a dynamic, open environment. The wireless channel changes frequently due to movements of objects or personnel and other wireless nodes are present using the same system. It is represented by a *reconfigurable discrete manufacturing system* consisting of flexible modules. The environment of the second category is almost static and is represented by a *wind energy plant* and its closed control loop for adjusting the blade pitch angle. Here, slight changes are only caused by the periodic rotation of the rotor itself, no other changes or movements occur.

Reconfigurable Manufacturing System

The first application category is represented by Reconfigurable Manufacturing Systems (RMSs), which are a concept addressing demand driven production processes [185], i. e., a consumer driven market that will demand customized goods in smaller production batches. Hence, flexible production lines are needed, that will be able to adapt quickly to a completely new manufacturing process. Reconfigurability is required at different levels and for different functionalities. In this work only the communication part of reconfigurability is considered by flexibly providing the needed network resources in a sufficient quality, which is also identified as an enabler for implementing the German strategic initiative *Industry 4.0* [86, 43].

The SmartFactory OWL (SFOWL)[1] is an example for a reconfigurable manufacturing system based on a modular mechatronic approach. The current architecture of the SFOWL is shown in Fig. 1.3 with a focus on flexible modules.

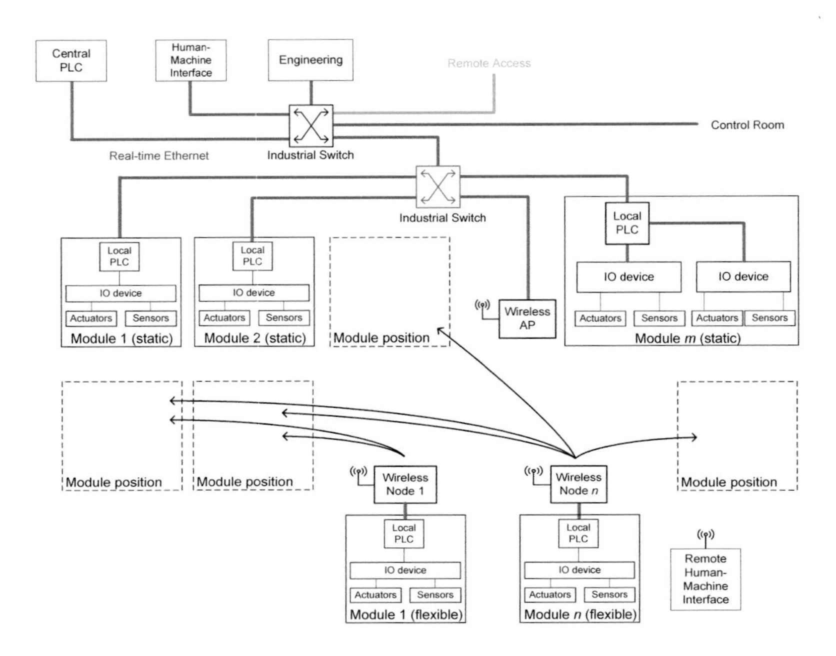

Fig. 1.3: Flexible modules within the SmartFactory OWL

It allows to add and remove certain production modules (depicted as *Module 1 (flexible) ... Module n (flexible)*) in order to adapt the overall manufacturing process. Each module is equipped with a local PLC, which uses a cyclic data exchange to acquire its local sensor values and to set its actuators, i. e., whenever a new module is added, it must be integrated into the whole system and therefore establish a real-time communication channel to a central PLC. A manufacturing system of this category is analyzed in Sect. 3.1 and it is found that within the modules a latency of $\leq$ 10 ms is required. The requirements for the communication between different modules can be classified into *real-time class 1* (cf. Table 1.1), because the time critical control loops are implemented within the module on the local PLC. In addition to this, such systems require several remote HMIs, which are equipped with standard Commercial-of-the-shelf (COTS) wireless hardware, for monitoring, and maintenance.

[1] `http://www.smartfactory-owl.de`

Wind Energy Automation

The second application category belongs to the area of renewable energies which is becoming more and more important in power supply systems. In Germany, it is expected that they will provide 47% of the overall energy consumption [11] in 2020. More than half of this capacity will be provided by wind energy plants. The market for automation solutions for wind energy will grow accordingly. For instance, an increase of the annual investments in this area by 80% is expected until 2020 [45]. Hence, industrial automation solutions for wind energy plants, for both on- and off-shore, are a tremendously growing field of application and gain an increasing relevance in industry.

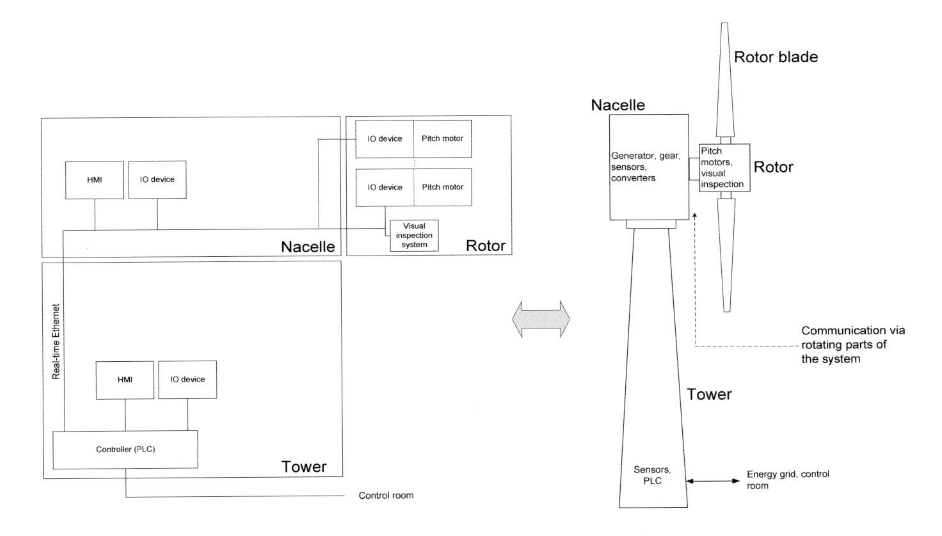

Fig. 1.4: Wind energy automation application

Wind turbines are very complex systems. Their internal automation architecture is shown in Fig. 1.4. Several different sensors and actuators are necessary for its control, e. g., to detect the wind direction, to detect the rotor position and adjust it accordingly, etc. Most of the available systems are variable speed wind turbines, which are controlled to optimize the generated power [44]. This is achieved by a closed-loop control system for adjusting the pitch angle. This is a challenging task, due to the rotation. The closed control loop of the pitch drives requires latencies of $\leq$ 10 ms and a jitter of $\leq$100 µs [84, 34], belonging to *real-time class 2*. Currently, the control loop is implemented by using wired fieldbuses with error prone and maintenance intensive slip rings to connect the controller in the nacelle and the sensors and actuators in the rotor. An automation device vendor, specialized for wind energy solutions, recently stated in [52] that it would be desirable to use wireless communication to connect the nacelle and the rotor, because of the very expensive slip rings, but existing wireless solutions are lacking the necessary determinism and reliability.

Besides controlling the wind turbine, several sensors are built-in for diagnostics. They are used to check whether all modules and system components are in an operational state, because downtimes of wind turbines are very costly, especially for off-shore systems: every maintenance case means that maintenance personnel must fly by helicopter to the turbine. In order to avoid such costly unplanned maintenance cases, the condition monitoring must also be improved for the blades and service technicians should be enabled to visually monitor the current condition of the blades online. To allow this, future systems require a video connection from the nacelle to the rotor resulting in a higher throughput demand.

Application Requirements

After analysing the application scenario including both application categories, several requirements are derived and must be fulfilled by the proposed concept. The requirements are grouped into *a*) temporal requirements, *b*) reliability requirements, and *c*) further functional/non-functional requirements. Formal definitions of the requirements and their relevant metrics are provided in Sect. 4.1.

The *temporal requirements* mainly determine the control-loop quality of an NCS. The temporal requirements of the application are timeliness and simultaneity. Timeliness is defined as the successful transmission of data within a given time bound, which is determined by the technical process. The relevant Quality of Service (QoS) parameter for timeliness is latency. The simultaneity requirement defines the maximum allowed deviation of consecutive cyclic data transmissions. The corresponding QoS parameter is jitter. Both requirements are quantified in terms of their metrics and separated into different real-time classes in Table 1.1. The described application categories belong both to *real-time class 2* and require the corresponding latency and jitter.

The *reliability* of the communication system is another important requirement for the envisioned application scenarios. It is mainly influenced by the wireless channel and its time varying behaviour. The reliability of the wireless communication system is defined as the Packet Loss Rate (PLR) perceived by the application. The resulting omission failures are tolerated by the application up to a given bound. If this boundary is exceeded, the application will not work as intended, e. g., resulting in a degraded manufacturing process. In the worst case, the manufacturing system or the wind energy plant is shutdown, which is very expensive and must be avoided under any circumstances. In the wind energy example, it might even cause damages of the system. If the communication is interrupted for a period longer than the *omission degree*, the system immediately goes into a safe state causing huge mechanical loads imposed on the tower. This is only possible for approximately ten times throughout the whole life time of the turbine. Both described application categories typically tolerate an *omission degree OD of 2*.

Further *functional and non-functional requirements* cover the remaining services and functionalities that are required for the application. Usually, a wireless network is integrated into an existing automation system as an extension of a wired communication network. In order to meet the challenging temporal requirements, an integration of our wireless solution into existing real-time Ethernet networks must be possible and both systems must allow a *synchronous operation*. Remote monitoring using visual inspection enables operators to diagnose specific parts of remote systems. For instance, when operating an off-shore wind energy plant down-times are extremely costly and must be avoided. A predictive maintenance using visual inspection of the blades requires a video stream from the rotor to the nacelle. Therefore, the communication system must provide a *sufficient capacity* for this kind of monitoring. Reconfigurable manufacturing systems are undergoing frequent changes of the plant due to *adding or removing new modules* in order to adapt the current technical process. The wireless communication system shall be therefore flexible enough to allow additional nodes to enter and to leave the system during its operational phase. Moreover, the concept must also *support the integration of other wireless nodes*, such as monitoring interfaces for operators as they are frequently used to maintain and operate the system. However, a degradation of the QoS parameters of real-time traffic due to the integration of such nodes can not be tolerated and must be avoided.

The identified requirements are summarized in Table 1.2 separated with respect to the application category.

Table 1.2: Requirements of both application categories

Requirement	**Reconfigurable manufacturing system**	**Wind energy automation**
Real-time class	1	2
Throughput capacity	≥ 1.5 Mbps Video stream for monitoring	≥ 1.5 Mbps Video stream for monitoring
System integration	Required	Required
Global time base	Required for synchronous operation	Required for synchronous operation
Standard nodes interoperability	Required for monitoring and maintenance nodes	Minor importance

1.2 Problem Exposition

Existing wired industrial Ethernet standards, for instance Profinet [60], are already able to fulfil the identified application requirements. After introducing them a few years ago, it is expected that they will be extensively used in future automation systems. This leads to a demand for a wireless counterpart to implement several applications, which is shown by the successful deployment of wireless technologies at different levels of the hierarchical pyramid. An example for the enterprise level is mobile access to the ERP system or a first prototype of a plant in its design phase which can be easily implemented neglecting the cabling. Wireless HMIs for monitoring and operation of plants are another example on the control level, which has already been deployed. However, all of these examples belong to diagnostics networks, as defined in Sect. 1.1. Neither temporal requirements nor reliability requirements are critical and must be met.

On the other hand, the NCS performance strongly depends on meeting these requirements. Unreliable communication networks degrade their behaviour severely and lead to instabilities. It may even result in equipment damage or even worse in personnel injuries. A few industrial wireless solutions for such control networks in factory automation are available. Industrial Wireless Sensor Networks (WSNs) are specifically tailored for usage at the field level [153, 94]. They are able to provide the required timeliness and have been explicitly developed for sensor and actuator communication. Thus, they are designed to transport only very small amounts of payload and their provided *capacity is limited.* Applications with a higher bandwidth demand, such as visual inspection and surveillance applications, are not supported by these solutions.

Another related approach are fault-tolerant industrial Wireless Mesh Networks (WMNs) [106], which provide a flexible infrastructure for control networks with application level end-to-end guarantees. They are able to achieve a high flexibility, due to the WMN, but a *trade-off regarding their temporal behaviour* must be made. The achievable guaranteed round trip times belong to real-time class 1 and can not satisfy the needs of the application scenario.

Due to its similarity to Real-time Ethernets (RTEs), the IEEE 802.11 [70] standard is another promising candidate for an industrial wireless communication system. It can provide the required capacity and can be transparently integrated into existing wired systems. However, IEEE 802.11-based Wireless Local Area Networks (WLANs) are not able to satisfy the application requirements in their current state, because of several identified reasons listed below:

The common distributed Medium Access Control (MAC) protocols are *non-deterministic in nature*, due to their backoff procedure and the uncoordinated, decentralized MAC protocol. This results in random latencies, and high latency jitter values, which must be avoided. They are very critical for the envisaged applications of NCS and cannot be tolerated by them. Even though coordinated mechanisms are provided, they turn out to be very *inefficient for frames with small payloads*, as usually found on the field level in industrial automation [42].

The shared wireless medium has a limited capacity and can be only used in half duplex mode. If its utilization increases in an uncontrolled manner, undesired overload situations might be the result and cause an *unpredictable behaviour* of the wireless network, which must be prevented. In this context, the interoperability with standard IEEE 802.11-based wireless nodes must be considered to avoid additional communication interferers.

Finally, a synchronous end-to-end delivery of application data is very important for achieving control loops having the expected quality for the second application category of wind energy plants. Existing wireless solutions are usually operated *asynchronously* and *totally decoupled* from the wired backbone. Considering typical industrial automation systems, a hybrid communication system of wired and wireless networks is very common [150]. Their asynchronous operation will decrease the achievable temporal behaviour significantly.

1.3 Solution Approach

In order to address the previously identified problems, this dissertation aims at designing an isochronous wireless network for industrial control applications. The system is based on IEEE 802.11 [70] to allow a seamless integration of the wireless system into existing RTEs and to ensure a sufficient capacity.

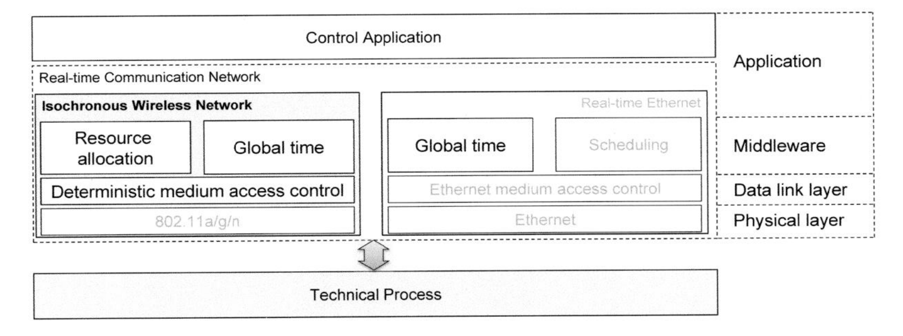

Fig. 1.5: Solution approach: Isochronous wireless network for industrial control

The solution approach is a flexible wireless real-time network for providing real-time communication services for Networked Control Systems. Several components of the approach are provided as integral parts to the system architecture of the European research project ^{flex}*WARE– Flexible Wireless Automation in Real-Time Environments* [163]. The solution approach is shown in Fig. 1.5. The approach consists of three main components: the medium access control, the resource allocation, and the provision of a global time base. In addition to this, the application, the industrial channel characteristics and the integration into existing wired networks are considered. All of these aspects are further described in the remainder of this section.

Deterministic medium access control is a very important characteristic of the system and guarantees a coordinated access to the wireless medium by means of using a Time Division Multiple Access (TDMA) based approach. It avoids random delays by using appropriate mechanisms on layer 2. The proposed MAC [164] is designed with a focus on transmitting small payloads [168] and, at the same time, supporting an adequate capacity for video surveillance applications. This can be achieved by using the existing WLAN physical layer and a modified MAC layer. A flexible flow control allows an adaptive retransmission handling to reduce the latency jitter. Furthermore, it is interoperable with existing wireless nodes and their operation does not influence the system.

Resource allocation is responsible for preventing overload situations in the wireless network and for providing mechanisms to guarantee that the real-time application requirements are met on the basis of admitted flows in the system. A cross layer frame inspection is used at the application interface to collect the application requirements, which are needed for the flow admission procedure [30]. The requirements are defined as tuples consisting of the relevant attributes, such as their send period and payload size. An online admission control in combination with an online scheduling allows new nodes to enter the system during operation [169]. The admission control is based on the specified requirements of a new flow, the current utilization of the system and the status of the wireless resource. It also considers priorities defined by the application to be able to reject flows with a lower priority. Once a new flow is admitted, the scheduling component adapts the assignment of time slots, i. e., the existing communication schedule, in accordance to the additional traffic flows without negatively influencing the existing flows. In order to optimize the communication jitter in the presence of frame errors, an adaptive retransmission handling following different policies depending on the application requirements is deployed.

Provision of a global time base allows the synchronization of all system components, i. e., all system components including the wired nodes rely on the same global time base. This avoids a significant deterioration of the temporal system behaviour by asynchronous communication processes. Besides guaranteeing a synchronous communication, a basic requirement for real-time communication systems, which rely on TDMA-based protocols, is that the nodes must share the same notion of time. For the sake of this coordinated access, which is not only limited to TDMA approaches, a common notion of time between the wireless nodes is mandatory. The global time base will be provided by a wireless clock synchronization system which relies on the Precision Time Protocol (PTP) [68] on the application layer [112]. Since this is also deployed in many RTEs, a simple integration of the wireless and wired clock synchronization is possible. The wireless clock synchronization is designed in a very efficient way, i. e., no additional overhead is introduced. This is a crucial property, especially in wireless networks with a limited capacity due to the shared medium.

Application and wireless channel pose additional constraints and challenges on the wireless communication system and are important to be considered. Both aspects have been analyzed in this work to derive valid application and channel models [83, 9]. The application mainly determines the requirements for our system, especially with respect to temporal behaviour. The characterization of the industrial wireless channel has a major influence on designing suitable mechanisms for the MAC and for the resource allocation.

1.4 Structure of the Thesis

The remainder of this thesis is structured as follows. In Chapter 2 related work and relevant background for this work is discussed and existing limitations are pointed out. The industrial environment is characterised in Chapter 3 in terms of industrial traffic characteristics and the industrial wireless channel. The approach of an isochronous wireless network for industrial automation and its integration into existing wired networks is discussed in Chapter 4. The following chapters describe the main concepts of the proposed solution approach in detail. In Chapter 5 a new medium access control protocol is presented and evaluated. The scheduling and admission control components for resource allocation are introduced in Chapter 6. Chapter 7 deals with all aspects related to the provision of a global time base. In Chapter 8 the prototypical implementation of the whole system and the implemented simulation model including the developed wireless channel model are described, followed by an evaluation of the overall system in two different case studies for both application categories. Finally, Chapter 9 will conclude this work and point out possible directions for future research.

2

Related Work and Background

This chapter provides a general state of the art and relevant background for this work. Both wired and wireless communication systems for industrial automation have been considered with their characteristics, even though the main focus is put on wireless systems. Existing technologies and solutions are analysed with respect to the relevant application requirements as identified in Sect. 1.1. The related work for the three main components of the solution approach, i. e., medium access control, resource allocation, and the provision of a global time base is discussed at the beginning of their corresponding chapter.

2.1 Communication in Industrial Automation

Nowadays, three generations of field level networks exists [152]. Even though traditional field bus systems are still used in many applications, future field level networks will be based on Real-time Ethernet (RTE), because of vertical integration aspects [151]. Hence, this section is focussed on existing real-time Ethernet protocols, and their characteristics. A specific focus is put on approaches based on TDMA, because they are able to provide the lowest latencies, i. e., the most powerful real-time communication features. Due to the fact that the wired system is not a research subject of this work, the considered solutions for the wired system are based on international standards and selected real-time Ethernet standards from IEC 61784-2 [63] are analysed for the real-time communication part.

2.1.1 Real-time Ethernet

Most of the existing RTE solutions are based on the IEEE 802.1 standard with switched Ethernet [65] and use the prioritization 802.1q [64]. However, depending on the required determinism, this might be not sufficient [78] and several extensions are developed which can be classified accordingly.

In general, RTE protocols can be also classified according to real-time classes (cf. Table 1.1). The classification and the differences of the protocol stack are both shown in Fig. 2.1. The protocols are becoming more powerful from class 1 to class 3, while additional functionality on the data link layer is introduced at the same time [77]. Modbus/TCP [58] and Ethernet/IP [61] are representatives of the first category. The protocols are using Ethernet as it is, only adding an industrial-automation specific application layer on top of TCP/IP. Because of using the whole TCP/IP protocol stack, their real-time characteristics suffer, and cyclic update times of about 100 ms can be typically achieved.

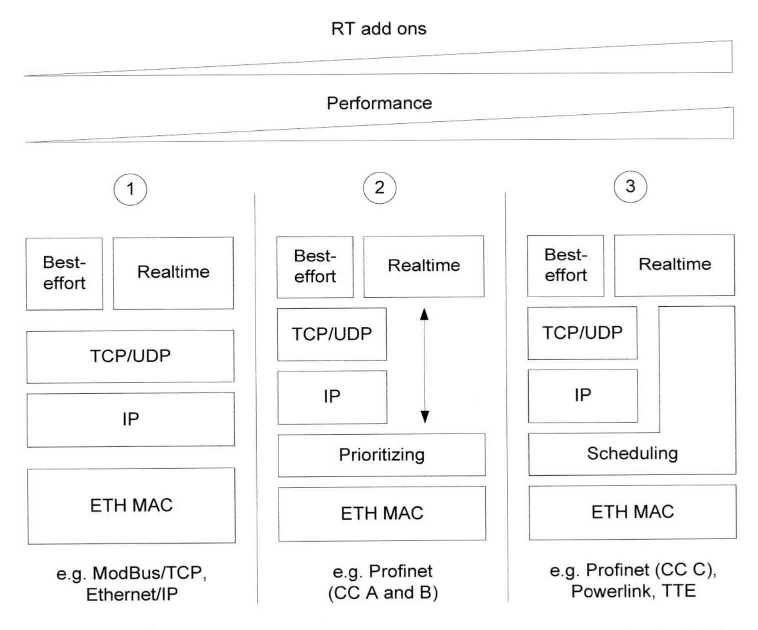

Fig. 2.1: Classification of Real-time Ethernet protocols [77]

Industrial communication systems, which use protocols of the second category, are perceived as a compromise between the native Ethernet standard and achievable real-time data transfer. An example for this category is given by Profinet (conformance class A and B) [60]. These protocols use a priority scheme at the Ethernet MAC layer. For an additional optimisation the transport and the network layer has to be bypassed for real-time data. Implementing this optimisation, cyclic update times in the range of $\leq$ 10 ms can be achieved.

Further enhancements are only possible by adapting the scheduling procedure of the MAC layer. Protocols which are changing the original MAC scheme of Ethernet are part of real-time class 3. In order to use those protocols, specific hardware or software is necessary. For example, Profinet (conformance class C) [60], Ethernet Powerlink [56], and Time-triggered Ethernet [92] belong to this category. Protocols which belong to class 3 are able to support cyclic update times in the range of $\leq$ 1 ms.

Common comparison criteria, also referred to as performance indicators, of the existing RTE communication standards are defined in [63] as communication profiles. It should be noted that all performance indicators highly dependent on the selected application scenario, i. e., payload, real-time class, chosen topology, etc. Relevant performance indicators for this work are briefly introduced in this section and summarized in a comparison table (cf. Table 2.1) followed by a description of the most relevant communication profiles.

Physical Layer The supported physical layers of the communication profiles are listed in the second column of Table 2.1. These can include IEEE 802.3 (Ethernet), IEEE 802.11 (wireless) and others. Ethernet is the common denominator of all communication profiles, even though always in a modified version for providing real-time.

Real-time mechanism The Real-time guarantee mechanism is used by a communication profile to guarantee an upper bound of the delivery time through the network. This is guaranteed in different ways. Frame prioritisation allows to send frames with a higher priority before frames with a low priority. It gives only moderate results as compared to the other two mechanisms. In the event-based/event triggered synchronization the communication partners synchronise their transmissions based on events generated by a defined master. This mechanism is usually referred to as polling. In TDMA all communication entities have a common schedule and their own time-slot for transmission. The mechanism requires synchronized distributed clocks with a high accuracy.

Communication model The communication model of each communication profile describes the method of organizing the communication on the application layer. The model is either a master/slave or a producer/consumer model. The *master/slave* or client/server model has at least one station which coordinates the communication by sending a request for data to its slaves, and the slaves reply accordingly. This model is considered as a confirmed service from the masters viewpoint and consists of the four service primitives request (master), indication (slave), response (slave), and confirmation (master). In the *producer/consumer* model, a producer (sender) sends data to one or more consumers (receivers). The data is sent either cyclically or event-based to the consumers. This model is considered as an unconfirmed service and consists of the two service primitives request (producer), and indication (consumer).

Delivery time The delivery time is the whole transmission time needed for a real-time Application Protocol Data Unit (APDU), i. e., from the sending station to the receiving station. It is measured at the application interface (cf. [63]). Therefore, a maximum delivery time or a range is given, indicating at which point the maximum value can be considered as a worst case. The delivery time depends on several factors, such as the cycle time, or the network topology (number of switches).

The delivery time can be considered as equivalent to the latency T_{Lat} defined in Sect. 4.1, which is used as a central metric for this work.

Network topologies Each communication profile supports a discrete set of network topologies which are star (hierarchical star), line, mesh (partial or full), and ring. Combining different topologies is also possible, for example, a tree topology is considered as a combination of a line and a star topology.

Table 2.1: Most common RTE profiles and their comparison criteria [63]

RTE Profile	Physical Layer	Real-Time mechanism	Communication model	Delivery Time	Network topologies
Ethernet/IP	IEEE 802.3	Prioritization	Producer/ Consumer	$< 20{,}4\,\mathrm{ms}$	Star Star
Profinet IO Conformance Class C	IEEE 802.3	TDMA	Producer/ Consumer	$\leq 1\,\mathrm{ms}$	Line, ring star
TTEthernet	IEEE 802.3	TDMA	–	–	Line, star
EtherCAT	IEEE 802.3, IEC 61158-2	Summation frame	Master/ Slave	$< 150\,\mu\mathrm{s}$	Line, star
ETHERNET Powerlink	IEEE 802.3	Polling mechanism	Master/ Slave	$< 1100\,\mu\mathrm{s}$	Line, star

Ethernet/IP

Ethernet/IP [61] is an application layer protocol which makes use of TCP/IP at the lower layers to transfer the information across the network. The physical layer is Ethernet and the MAC layer uses CSMA/CD for channel sensing. Because of such a channel access scheme, deterministic communication across the whole network is impossible. There are two profiles which are available for Ethernet/IP. The first profile is based on frame prioritisation and does not need special hardware on the end-stations or the switches. In this profile, the frames are prioritised and the scheduling algorithm allows a faster processing of high priority messages. In the second profile, the clocks of the devices are synchronized to each other with IEEE 1588v2 [68] and allow a scheduling of messages. Hence, the communication becomes more deterministic. The cyclic real-time frames between a producer and other consumers are tagged as *implicit* and transferred through UDP/IP.

Profinet IO

Profinet IO [60] is currently the most important view for Profinet. It is allowing real-time communication and addresses decentralized field devices. Profinet IO has three conformance classes supporting different real-time requirements. Conformance class C, formerly referred to as Profinet IRT (Isochronous RT), is specifically designed for

NCS, and supports real-time constraints. In Profinet the standard MAC of Ethernet is modified and the medium access is based on TDMA. The whole real-time data traffic is scheduled a priori based on different requirements of the application. The nodes share a common notion of time which is obtained by the precision time control protocol (PTCP), a modification of IEEE 1588v2 PTP [68].

Time-triggered Ethernet

The TTEthernet system [92] uses the same infrastructure for real-time communication and best-effort traffic by adding a TDMA scheme to a switched Ethernet system. All TTEthernet share a common schedule which is provided by a global scheduler and transferred to all TTEthernet nodes and switches. The clock synchronization for the global time is based on IEEE 1588v2 [68]. The protocol supports time-triggered traffic as well as rate constraint traffic and best-effort traffic. Rate constraint traffic is used for applications with reduced real-time requirements.

Ethercat

Ethercat [55] is a master/slave based system and uses a summation frame for transmitting real-time critical process data. The network is segmented into Ethercat segments. The Ethercat master sends cyclic frames to its slaves. Each slave processes the received frame on-the-fly and forwards it to the next slave. Ethernet frames are transmitted through all slaves on the segment, in sequence, from the first slave device (the owner of the MAC address of the segment, or the segment address slave) to the last slave device. No intermediate switches between master and slave are used in Ethercat, since every slave is a 2-port switch. However, the slaves require special hardware to support the on-the-fly frame processing. The profiles of Ethercat are based on an event-based synchronization and a clock synchronization scheme, the second one allowing for real-time communication.

Ethernet Powerlink

Ethernet Powerlink [56] is based on a master/slave communication model on a shared Ethernet segment. The master, which is also called managing node, coordinates the communication with every slave and ensures deterministic real-time communication. Four phases are forming a cycle, start, cyclic, asynchronous, and idle. In the start phase the master multicasts the start of cycle (SoC) message to start the cyclic phase. In this phase the master polls each slave separately and gets a reply from the slaves. After polling the active nodes, there are some slots reserved for asynchronous communication where non real-time traffic is sent followed by the idle phase until the next cycle. There are no restrictions on which topologies Ethernet Powerlink when using hubs. Even though switched Ethernet can also be used, it is not recommended. Switches will cause an additional jitter and path delay which reduces the achievable performance of the protocol.

2.1.2 Engineering Aspects and Flexibility

The commissioning of today's Real-time Ethernet (RTE) systems requires a time consuming and error-prone manual system configuration process. The corresponding common engineering cycle is always based on a static offline configuration phase as discussed in [33]. This is due to the need to maintain a high level of determinism for the production process and avoiding costly production interruptions. It consists of three steps, (i) the control application is implemented in an engineering tool, (ii) the physical structure of the automation system, including all devices, is added to the engineering tool and configured accordingly, and (iii) The logical variables of the control software are mapped to the physical sensor and actuator data. Only the second step is further considered here.

The procedure for the second step is as follows. The device information of all IOD of the system, referred to as Device Description (DD) files, is transferred offline to an engineering tool. The DD file is then used to create a project with configuration and parameterisation as required by the application. In this step the parameters and the communication schedule is calculated for all devices of the system, even the ones that are inactive. Afterwards, it is passed from the engineering to the PLC which configures all available IO Devices (IODs) in the start-up phase of the system.

Hence, the existing systems are very static and flexibility as required for the first application category of RMS must be engineered a priori. Interesting solution approaches are discussed by Dürkop et al. in [32] for the configuration and by Wisniewski et al. in [183] for scheduling, but not further considered in this work.

2.2 Industrial Wireless Communication

Typical application areas for industrial wireless networks can be found in the Process Automation (PA) as well as in the Factory Automation (FA) domains. Building automation is also relevant, but not further considered in the context of this work. Both domains have different requirements [172]. Wireless systems in the PA domain are mainly used for monitoring, process control and asset management. The wireless technology has to span distances of more than 1 km, but with moderate latencies. Whereas applications in the FA domain might even deal with control loops, i. e., stationary and mobile sensors, and monitoring. In many applications of FA you have to cover a limited range (e. g., one single manufacturing cell), but the temporal requirements are in the range of real-time class 2 and very challenging. This section provides an analysis of existing industrial wireless solutions and their characteristics based on the identified requirements of the application categories (cf. Table 1.2).

The existing wireless solutions for industrial automation are based on COTS technologies, due to economical reasons and a wider market acceptance [39]. In FA existing IEEE 802.11 and IEEE 802.15.1 conform components are typically used with a few proprietary protocol extensions. For instance, the iWLAN system of Siemens [157] or the Wireless Interface for Sensors and Actuators (WISA) system of

ABB [153]. Another solution is the Wireless Sensor and Actuator Network (WSAN) for factory automation [94], which is specified by the Profibus and Profinet user organization (PNO). Its fundamental concepts are based on WISA, but it has been extended with a few modifications.

In the PA domain, system based on IEEE 802.15.4 are used due to their low energy consumption. Based on this, the standards WirelessHART and ISA 100.11a are developed by the HART communication foundation and the international society of automation (ISA), respectively.

2.2.1 Systems Based on IEEE 802.15.1

The IEEE 802.15.1 standard [66] specifies the Physical Layer (PHY) and the Medium Access Control Sublayer (MAC) of the well-known Bluetooth (BT) technology and it operates in the 2,4 GHz-ISM-Band. A typical application for the 802.15.1 would be for instance a connection between a cell-phone and a headset or a headphone and an audio device. However, the system is also attractive for the use in industry applications [181], since its medium access method allows a deterministic data delivery.

According to 802.15.1 every physical channel used, is called a piconet consisting of one master, and up to seven slaves. The master device within a piconet coordinates the traffic within a piconet and starts the connection setup procedure. Furthermore, the wired and the wireless network are connected via the master device. The other devices are called slave devices. These are only able to directly communicate with their master. Slave devices within a piconet are either in an active or inactive operation mode. An active slave device communicates with the master device, inactive devices are in an energy saving state and only start a communication upon a wake up call by the master device.

Piconet traffic is based on TDMA with a duplex scheme, allowing the master to communicate in odd-numbered time slot. The slaves are only permitted to reply in time slots which are even-numbered and after they have been polled by the master. The channel access is divided into slots, each of 625 µs length. A data transmission may occupy the channel for 1, 3 or 5 consecutive slots. A frequency hopping spread spectrum (FHSS) modulation with 79 different 1 MHz channels is used to deal with the ISM-Band requirements. In order to avoid problems with the coexistence to other technologies, an adaptive FH (AFH) can be used to render occupied frequencies.

The Data Link Layer protocol is mainly divided into asynchronous connectionless (ACL) and Synchronous Connection Oriented (SCO) frames. The SCO transmission is attractive for industrial usage. The SCO transport mechanism reserves time slots on the physical channel. It is therefore able to provide QoS guarantees. The transmission of an SCO packet takes 366 μs and typically transmits data with a rate of 64kb/s [181]. This is done by reserving periodic slots for a specific communication between the master and a slave device. An extension to the SCO data transmission is the eSCO data transmission. It also uses reserved time slots and has the advantage that it can deal with different, but still static, transmission rates and is able to transport non voice data [66]. Nonetheless, in most cases, it is necessary to have more than one piconet, because of the existing slave limitation to seven.

Wireless Interface for Sensors and Actuators

The WISA system [153, 90] has been developed by ABB and is targeted at typical factory automation applications. It has been designed to cover wireless communication as well as wireless power supply. This enables truly wireless connected sensors and actuators without having the need for a seperate power supply.

The wireless communication WISA COM is based on IEEE 802.15.1 physical layer. In a system that needs to achieve the delivery of messages with a very high probability of success and high number of devices, the medium access is important. The medium access in WISA is time division multiple access with frequency division duplex and frequency hopping (TDMA/FDD/FH). The WISA frequency hopping scheme guarantees that the frequencies used in successive frames are widely spread. The downlink transmission from the base station to the wireless devices is always active, for the purpose of establishing frame and slot synchronization for the devices, but also to send acknowledgements and data.

In order to save power, uplink transmissions from a sensor only occur when it has data to send. The requirement of wireless real-time communication combined with a need for a high number and high density of devices, makes efficient use of the available bandwidth very important. A number of input modules (base stations) can be distributed in the plant, with short-range communication to local sensors/actuators. The input modules (base stations) are connected to the control network via any field bus and communicate with the local wireless devices. A sophisticated input module (base station) ensures that the complexity resides in the input module rather than in the wireless sensors or actuators. One input module can handle up to 120 devices (sensors) or 13 wireless sensor input/output pads. Since in typical applications only sensor and actuator devices are targeted the capacity of the system is very limited. It is possible to exchange only a few bytes of payload.

Wireless Sensor and Actuator Network for Factory Automation

The WSAN for factory automation is developed by the Profibus User Organization and is partly based on the previously introduced WISA system. It also uses the PHY layer of IEEE 802.15.1 and provides a frequency hopping multiple access, i. e., a combination of frequency hopping and TDMA. The whole system is specifically designed for the sensors and actuators on the field level. WSAN is able to address up to 120 wireless nodes in the system, targeting applications of real-time class 2. The reliability is achieved by allowing up to four retransmissions on different frequencies. In this case the update time must be decreased. Moreover, a blacklisting of certain channels is provided to minimise interference with other system. However, the capacity of the system is limited due to the same reason as for WISA.

2.2.2 Systems Based on IEEE 802.15.4

The IEEE 802.15.4 standard [69] has become a communication standard for low data rate, low power consumption and low cost Wireless Personal Area Network (LR-WPAN). The protocol focuses on very low cost communication, which requires very little or no underlying infrastructure. The basic framework supports a communication range of $\leq$ 10 m. The capacity of the system varies depending on the selected data rate of 20, 40, 100, and 250 kbps. The protocol provides flexibility for a wide variety of applications by effectively modifying its parameters. It also provides real-time guarantees by using a Guaranteed Time Slot (GTS) mechanism for time sensitive applications. Hence, two kinds of network configuration modes are provided in [69]. The beacon enabled mode, where a PAN Coordinator periodically generates beacon frames after every Beacon Interval (BI). In the non beacon enabled mode, all nodes can send their data by using an unslotted CSMA/CA mechanism which does not provide any time guarantees to deliver data frames.

The Physical Layer (PHY) is responsible for transmission and reception of data using a selected radio channel according to the defined modulation and spreading techniques. The spreading in all frequency bands is based on Direct Sequence Spread Spectrum (DSSS). The different modulation schemes are Binary Phase Shift Keying (BPSK), Amplitude Phase Shift Keying (ASK) and Offset Quaternary Phase Shift Keying (QPSK). The choice of a modulation scheme depends on the desired data rate.

The IEEE 802.15.4 MAC layer provides features like: beacon management, channel access, GTS management, frame validation, acknowledgment frame delivery, association, and disassociation. In the beacon enabled mode, the PAN Coordinator uses a superframe structure in order to manage the communication between its associated nodes. The superframe structure is defined by means of two parameters: *Beacon Order (BO)* and *Superframe Order (SO)*. The active period (superframe duration (SD)) is divided into 16 equally sized time slots for data transmission. Within the active period two medium access coordination functions are defined in IEEE 802.15.4: a mandatory Carrier Sense Multiple Access (CSMA) mechanism for the contention access period and an optional GTS mechanism for the Contention-free Period (CFP). The contention access phase shall start immediately after the beacon and complete before the beginning of the CFP on a superframe slot boundary. The CFP shall start on a slot boundary immediately after the CAP and it shall complete before the end of the active period of the superframe. The CFP can be activated by a request sent from a node to the PAN Coordinator.

1. *CSMA:* Two versions of CSMA/CA are defined, the unslotted for the non beacon-enabled mode and the slotted CSMA/CA for the beacon-enabled mode. For both versions it is based on backoff periods.
2. *GTS:* GTS provides real time guarantees for time sensitive applications. GTS can be activated by the request sent from a node to the PAN Coordinator. At the reception of this request, the PAN Coordinator checks whether there are sufficient resources available for the requested node in order to allocate requested time slot. Maximum of 7 GTSs can be allocated in one superframe.

Existing industrial solutions are described in the next two subsections. They are based on IEEE 802.15.4 and mainly targeting applications of the process automation domain and building automation, because during designing the IEEE 802.15.4 standard the focus was put on a very low power consumption of the nodes rather than on their real-time capabilities [93, 51].

WirelessHART and ISA 100.11a

WirelessHART and ISA 100.11a are quite similar with only a few differences which are not relevant for this work (cf. [140]). Therefore, only Wireless HART is introduced here, since it is accepted as international standard IEC 62591 [59] since 2010 and gains a wider acceptance at the moment. WirelessHART [59] can be considered as the wireless extension of the Highway Addressable Remote Transducer (HART) protocol extensively deployed in process automation. The system is based on the 802.15.4 standard with several modifications of the MAC layer, a network layer, a transport layer and the HART application layer. The MAC is a combination of TDMA with a synchronized frequency hopping called time synchronized mesh protocol (TSMP). It defines time slots of 10 ms which are grouped into periodic superframes. The system is deployed in a mesh topology and gateways provide the access to the plant backbone network. The scheduling assigns different receive and transmit timeslots to individual devices and along certain routes in the network. The scheduling is done by a centralized network manager which is the core component of the wireless system and also responsible for a continuous monitoring and adaptation if necessary. Unfortunately, for both systems the network manager is not part of the standard and depends on proprietary implementations. The minimum achievable latency for both systems is approx. 100 ms but depends on the application [140].

Time Synchronized Channel Hopping and 6TiSCH

Recently, the IEEE 802.15.4e [71] standard was introduced as an amendment of IEEE 802.15.4 which is targeting industrial applications. Its main features are the time synchronized channel hopping (TSCH) to increase the reliability of transmissions, being similar to TSMP, and reduce the energy consumption of nodes. TSCH is based on time slots which are typically 10 ms and grouped into one slot frame. The slotframe is periodically repeated. The mechanism requires that all nodes in the network are synchronized and adhere to a schedule which defines the time of transmission, time of reception, sleep time, etc. for each time slot. The mechanism also defines a channel offset for each slot which is used to calculate different transmission frequencies for consecutive slotframes resulting in a channel hopping. Due to the mesh topology of the network, the achievable latency is in the order of $\geq$ 100 ms [125]. The required accuracy of the clocks is in the range of 1 ms [159].

In this context the 6TiSCH activities within a new Internet Engineering Task Force (IETF) working group must be considered [134]. 6TiSCH is working on an architecture to use IPv6 in 802.15.4e based networks using an operation sublayer

(6top). This sublayer will be mainly responsible for the scheduling of data transmissions, i. e., the resource allocation, because 802.15.4e is only specifying the MAC but not the resource allocation itself.

2.3 IEEE 802.11

The IEEE 802.11 standard [70], generally referred to as WLAN, is part of the IEEE 802 standards family. The comprised standards fall within the scope of layer one and layer two of the ISO/OSI reference model and specify the data link layer in two sublayers, the Logical Link Control (LLC) and MAC. They have some standards within the data link layer in common, such as IEEE 802.2 for the LLC layer and the IEEE 802.1 standards for bridging, but they differ significantly when it comes down to the physical layer (PHY) and some other parts of the MAC sublayer. The main direction of this section is towards the specified medium access control and its analysis, but first a brief overview of the physical layer is provided.

2.3.1 Physical Layer

The original 802.11, dating back to 1997, supports three different options of a PHY layer, which are an optical Infrared (IR) PHY layer, a Frequency Hopping Spread Spectrum (FHSS) layer and a Direct Sequence Spread Spectrum (DSSS) layer, all supporting basic data rates of 1 and 2 Mbps. In addition to this further high rate extensions were developed. IEEE 802.11b operates in the 2,4 GHz-ISM-band for data rates up to 11 Mbps still using DSSS. IEEE 802.11a in the 5 GHz-ISM-band for max. data rates of 54 Mbps based on the Orthogonal Frequency Division Multiplex (OFDM) technology. 802.11g was specified to allow data rates of 54 Mbps in the 2,4 GHz-ISM-band, is also based on OFDM and backwards compatible with 802.11b. The 802.11n extension was recently added. It allows a further increase of the capacity up to 600 Mbps by using multiple independent channels, referred to as Multiple Input Multiple Output (MIMO), as well as mechanisms to aggregate data and channel bonding to increase the available channel bandwidth. The available data rates, spreading technologies and modulation types are shown in Table 2.2 for the most common standard amendments.

In today's 802.11 systems only DSSS and Orthogonal Frequency Division Multiplexing (OFDM) based modulation types are relevant. In the Direct Sequence Spread Spectrum Signaling (DSSS), the data symbols are spread over the full bandwidth (spectrum) of a device's transmitting frequency. The data signal at the sending station is combined with a higher data rate bit sequence, or chipping code, that divides the user data according to a spreading ratio. The chipping code is a redundant bit pattern for each bit that is transmitted, which increases the signal's resistance to interference. OFDM is a Frequency-Division Multiplexing (FDM) scheme utilized as a digital multi-carrier modulation method. OFDM splits the radio signal into multiple smaller sub-signals that are then transmitted simultaneously at different frequencies to the receiver. OFDM is used in all recent

Table 2.2: WLAN 802.11 modulation types and data rates

Standard (Release)	**ISM-band** (GHz)	**Max. data rate** (Mbps)	**Modulation types**	**FEC** (code rate)
802.11a (1999)	5.0	54	OFDM (BPSK, QPSK,16-QAM, 64-QAM)	1/2, 2/3, 3/4
802.11b (1999)	2.4	11	DSSS, CCK (BPSK, QPSK)	–
802.11g (2003)	2.4	54	OFDM (BPSK, QPSK, 16-QAM, 64-QAM)	1/2, 2/3, 3/4
802.11n (2009)	2.4 / 5.0	600	OFDM (BPSK, QPSK, 16-QAM, 64-QAM)	1/2, 2/3, 3/4, 5/6
802.11ac (2014)	5.0	1600	OFDM (BPSK, QPSK, 16-QAM, 64-QAM 256-QAM)	1/2, 2/3, 3/4, 5/6

standard amendment 802.11a/g/n/ac due to the increased robustness especially in environments with many interferers [147]. It can be used with different modulation schemes and Forward Error Correction (FEC) code rates ranging from BPSK with a code rate of 1/2 to 256-QAM (quadrature amplitude modulation) with a code rate of 5/6.

2.3.2 Medium Access Control

According to the original IEEE 802.11 MAC protocol [166], the architecture of the MAC sublayer includes a mandatory Distributed Coordination Function (DCF) and an optional Point Coordination Function (PCF). However, due to the limitations of this architecture for transmitting time critical traffic flows, real-time enhancements were introduced. They are addressed by the standard amendment IEEE 802.11e which has been included into the standard in 2007. It provides advanced QoS capabilities by adding the Hybrid Coordination Function (HCF), which defines two new access mechanisms.

The DCF is totally distributed and can be used within both ad-hoc and infrastructure network configurations. Conversely, the PCF is centralized and can be used only on infrastructure network configurations. In the DCF, each station senses the shared channel and transmits when it finds a free channel according to a carrier sense mechanism. In the PCF, a coordinator station polls the other nodes thus enabling them to transmit in a collision-free way. The Enhanced Distributed Channel Access (EDCA) is based on the DCF medium access, except of the ability to assign different priorities to various kinds of traffic flows. The HCF Controlled Channel Access (HCCA) defines a parameterized QoS support. It is also relying on a polling procedure with an underlying TDMA principle, but more flexible as compared to the legacy PCF. Currently, the most widely used channel access mechanisms are DCF and EDCA, while the PCF and the HCCA have received little attention so far. Especially as far as commercial products are concerned, because available chipsets are still lacking support for these mechanisms. In the remainder of this subsection, the four different access methods are described.

Distributed Coordination Function

The DCF is based on a Carrier Sense Multiple Access mechanism with Collision Avoidance (CSMA/CA) and works as follows: first it determines whether the medium is idle or busy, as soon as the idle state is detected, the station waits for an DCF Interframe Space (DIFS). If the channel remains idle, a backoff procedure starts. A backoff timer is set to a number randomly chosen out of a contention window CW and then decremented periodically. When it reaches zero the station starts to transmit its pending data frame. However, no QoS is supported by this function, since no means to differentiate the frames in terms of different priorities are provided.

According to the IEEE 802.11 MAC protocol, an Acknowledgment (ACK) frame has to be sent by the receiver to notify the transmitter about a successful reception of a data frame. The time interval between the reception of a data frame and the transmission of the relevant ACK is defined as Short Interframe Space (SIFS). This small gap between transmissions gives priority to this frame exchange sequence, by preventing other nodes from accessing the channel, because they have to wait for the medium to be idle for a longer time interval, at least for a DIFS.

In 802.11 physical and virtual carrier-sense functions are used to determine the state of the medium. Either function is able to indicate a busy medium. The physical carrier-sense mechanism is provided by the 802.11 PHY, while the virtual carrier-sense mechanism is provided by the MAC and is referred to as the Network Allocation Vector (NAV).

Point Coordination Function

The central component of the PCF is a Point Coordinator (PC). It is acting as the polling master, which determines which station is currently allowed to transmit. The operation of the PCF may require additional coordination, that is not specified in the standard, to permit efficient operation in cases where multiple point-coordinated networks are operating on the same channel and share an overlapping physical space.

The PCF uses a virtual carrier-sense mechanism aided by an access priority mechanism. The PCF distributes information using Beacon management frames to gain control of the medium by announcing the start of a CFP and setting the NAV in the stations. In addition, all frame transmissions under the PCF may use a PCF Interframe Space (PIFS) that is smaller than the DIFS for frames transmitted via the DCF. This means that point-coordinated traffic has priority in accessing the medium over stations in overlapping networks operating in the DCF mode. Such an access priority may be utilized to realize a contention-free access method. The PC controls the frame transmissions of the stations, so as to eliminate contention for a limited period of time. Although the ability of providing support for collision-free transmissions would be very beneficial to industrial communication, especially when handling time-constrained traffic, PCF has not been implemented in real devices, because of lacking support from chipset vendors.

Enhanced Distributed Channel Access

The EDCA is an extension of the DCF and defines eight different priority levels. The priorities are mapped into four Access Categories (AC) in compliance with the IEEE 802.1D standard [65] thereby providing differentiated and distributed channel access. Within a QoS enabled wireless station (QSTA) and a QoS enabled (QAP) every AC is represented as an independent transmission queue. Every single queue contends for the medium separately and has different parameter sets for accessing the channel.

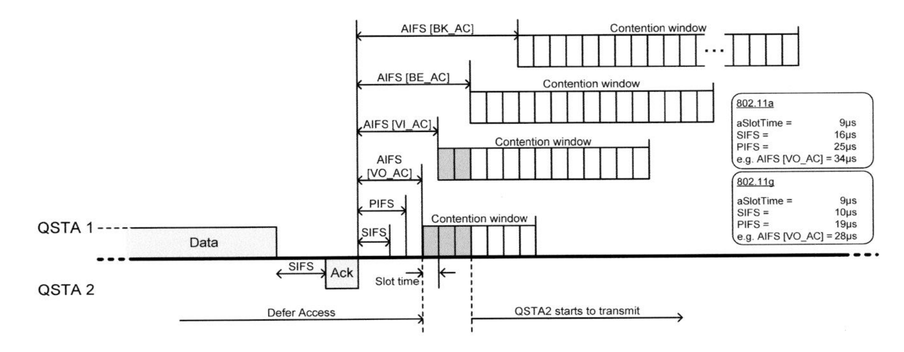

Fig. 2.2: IEEE 802.11 channel access with different priorities

The parameter set consists of the Arbitration Interframe Space number (AIFSN), the two bounding values for the contention window CW_{min}[AC] and CW_{max}[AC] and the maximum allowed transmission time for one station $TXOP_{limit}$. The AIFSN depends on the AC and is used to derive the Arbitration Interframe Space (AIFS) with Eq. (2.1), where $aSlotTime$ and SIFS depend on the used physical layer (e. g., 9 µs and 16 µs for 802.11a or 9 µs and 10 µs for 802.11g).

$$\text{AIFS[AC]} = \text{AIFSN[AC]} \cdot aSlotTime + \text{SIFS} \tag{2.1}$$

Similar to the DCF, a station always has to wait an AIFS before it can contend for the medium an. The backoff procedure is also similar to the DCF, except of different values for CW_{min} and CW_{max}. The contention for the medium with different priorities is shown in Fig. 2.2. As a result, the EDCA parameter sets of the four ACs cause different waiting times, i. e., a decreased waiting time for high priority frames and a longer waiting time for low priority frames. Therefore the probability of a successful data transmission of the high priority traffic is increased.

HCF Controlled Channel Access

The contention-free medium access in 802.11 [70] is realized with the HCF Controlled Channel Access (HCCA). The HCCA is the replacement of the previously defined PCF and used for controlled channel access. The time between two consecutive beacon frames is called a *superframe*. A superframe is divided into an optional CFP and a Contention Phase (CP). During the CFP the Hybrid Coordinator (HC) controls the access to the channel by polling its associated stations with QoS requirements. Besides, the HC is allowed to initiate a Controlled Access Phase (CAP) during the CP after detecting that the channel is idle for a time interval longer than a PIFS (PIFS = SIFS + $aSlotTime$) and whenever there is a need to transfer time critical data. An example for a superframe is shown in Fig. 2.3. A polled station is granted a Transmission Opportunity (TXOP) allowing the station to occupy the channel for a time period equal to the TXOP value. This concept elevates the HCCA with greater flexibility than its predecessor, although the time for generating CAPs is limited to a maximum duration in order to leave space for stations operating under the EDCA.

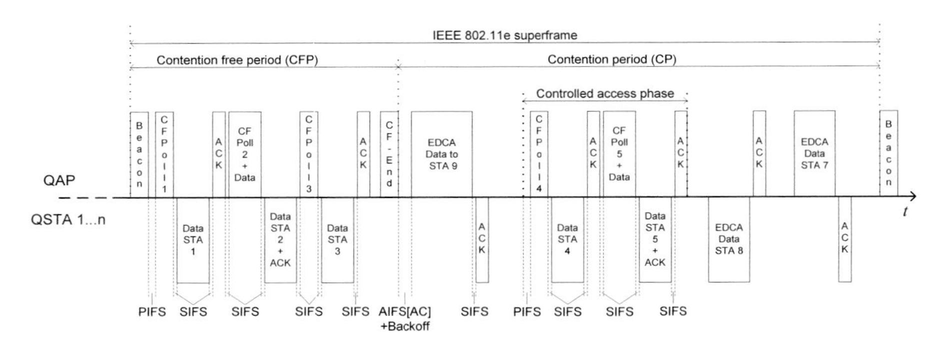

Fig. 2.3: Example of an IEEE 802.11 superframe

2.3.3 Admission Control and Scheduling

A very important concept for QoS support in 802.11e is the *admission control*. Whenever a station wants to associate with a certain Basic Service Set (BSS) it has to specify its requirements during a Traffic Specification (TSPEC) negotiation as it was introduced in the integrated services architecture [74]. The negotiation is done using TSPEC frames which may contain parameters related to the time critical traffic flow, such as *mean data rate* or *delay bound*. They are exchanged between the QAP and the QSTAs to establish a Traffic Stream (TS). The admission or rejection of the new TS depends on the adherence of its requirements. Every station can have up to eight different TS with different QoS requirements. In order to be enable the HC to compute a schedule three mandatory parameters are necessary, namely the Nominal MSDU Size (L), the Mean Data Rate (ρ) and either the delay bound (D) or the maximum service interval (MSI).

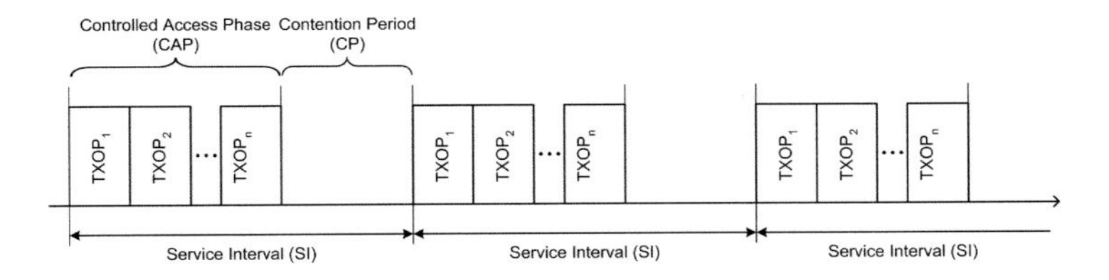

Fig. 2.4: A sample TXOP schedule for HCCA

After a successful negotiation, the HC calculates the new polling-schedule for all admitted stations based on the specified requirements. Three mandatory parameters are necessary, the Nominal MSDU Size (L), the Mean Data Rate (ρ) and the maximum service interval (MSI). The simple scheduler uses the negotiated parameters and schedules TXOPs for each station with a fixed length and at constant intervals. First of all, it calculates the Service Interval (SI) for successive TXOPs for the same station. It takes the minimum of all $MSIs$ for all TSs and chooses the SI lower than this value and as a sub-multiple of the beacon interval. Then the number of arriving frames (N) during one SI is computed with Eq. (2.2) for every traffic stream i of a station.

$$N_i = \left\lceil \frac{SI \cdot \rho_i}{L_i} \right\rceil \quad i = 1 \dots n \tag{2.2}$$

After this the TXOP duration for this particular station j is calculated with Eq. (2.3). It is the sum of the maximum of the time to transmit N_i frames at the PHY rate R_i and the time to transmit one maximal MSDU at R_i, both plus the additional overhead O for MAC and PHY header.

$$TXOP_j = \sum_{i=1}^{n} \max\left(\frac{N_i \cdot L_i}{R_i} + O, \frac{M_i}{R_i} + O\right) \tag{2.3}$$

Figure 2.4 illustrates an allocation schedule of the simple scheduler, which can be obtained with Eqs. (2.2) and (2.3). It consists of CAPs that are constructed by concatenating TXOPs from all stations. The CAP is always repeated with a period equal to the SI value and a length of the sum of all TXOPs. This simple scheduling algorithm will be used later on as a benchmark for our proposal, because it is a recommendation of the standard [70].

2.3.4 Relevant Amendments

Several other amendments exist which are dealing with improvements of the standard. However, only some of them are relevant for this work. A brief description of these additional features follows.

Dynamic Link Protocol and Block Acknowledgment

In order to decrease the protocol overhead IEEE 802.11 defines a way to directly communicate with other clients in infrastructure networks and to acknowledge more than one frame with a block acknowledgement. The direct link setup (DLS) enables two stations belonging to the same BSS to communicate directly to each other without directing the frames through the AP. Before the transmission starts, the direct link has to be setup via a request to the AP which is forwarded to the intended other station.

The block acknowledgement is also an optional feature which increases the throughput efficiency. With this option enabled, a station is allowed to transmit several frames within one TXOP. All transmitted frames form a single block which is acknowledged by only one block acknowledgement frame in the end, leading to a reduction of necessary control message exchanges.

Wireless Network Management

The IEEE 802.11v amendment enhances management services in the WLAN standard and has been integrated into the latest version of IEEE 802.11-2012 [70]. For instance, it will enable simplified wireless network management and will increase the scope of control for network administrators. The 802.11v improvements include procedures for wireless client control, which have influence on optimal working conditions of the network infrastructure. This standard will be responsible for load balancing, where mobile clients will be diffused between different APs. Another important aspect is the extension of clock synchronization features, because it specifies timing measurement frames and the inclusion of additional time information.

Robust Audio Video Transmission

Within the standard amendment IEEE 802.11aa [72], extensions for a robust and reliable transmission of audio and video data are defined. Furthermore, it promises to provide a full compatibility to the relevant mechanisms of IEEE 802.1AVB (802.1Qat, 802.1Qav, 802.1AS), which has been moved into the IEEE 802.1 TSN task group. Especially the stream reservation protocol (SRP) and its procedures for resource reservation is of interest, i. e., additional control frames are defined to allow an Access Point (AP) to act as intermediate node in a path between a talker and a listener [115]. Moreover, IEEE 802.11aa allows to prioritize between different video transport streams belonging to the same EDCA access category. The amendment explicitly addresses a better link reliability and low jitter characteristics by defining different policies for a group transmission service, but only for broadcast and multicast frames. Hence, the mentioned improvements are not relevant for this work, because the most relevant type of traffic within the defined application scenarios is unicast.

Security

It is well-known that security is of utmost importance in every industrial system, wired and wireless. Wireless systems are even more critical due to the shared medium with no access restrictions. However, it is sufficient for this work to apply the available security measures provided by IEEE 802.11, no further extensions are necessary. For a later deployment of this solution it is recommended to follow the procedural approach proposed by Treytl et al. in [161].

2.3.5 Industrial Extensions

The *Scalance W series* [157] is a range of wireless components based on IEEE 802.11, also referred to as *industrial WLAN (iWLAN)*, which have been developed by Siemens for industrial wireless communication. In order to allow a deterministic data communication, an industrial version of the point coordination function, called *iPCF*, was specified and implemented in their wireless components. The *iPCF* mechanism has been modified as compared to the standard PCF. For instance, the DCF is not supported and every transmission is based on polling [149]. Hence, it is a proprietary solution and not interoperable with standard IEEE 802.11 devices from other vendors. The *iPCF* relies on a central node, usually the AP, which cyclically polls all other associated nodes. The polling avoids the typical contention for the shared medium and a possible backoff procedure leading to a cyclic deterministic data transfer with network update times (NUT) $T_{NUT} \geq 16\,\text{ms}$. In order to minimise interferences with other wireless systems, it is recommended to deploy the *Scalance W* system with special radiating cables as antennas. These antennas provide a precisely defined coverage area, but they are limiting the mobility of nodes to positions in close proximity along the installed cable.

Even though the *iPCF* provides a coordinated medium access and allows only one node at a time to transmit data, the system is operated asynchronously to existing RTEs and no global time base is established. Moreover, the system is not interoperable with standard WLAN nodes which could result in additional disturbances from standard nodes.

To the best of the authors knowledge [28], all other existing 802.11-based industrial solutions are using standard EDCA allowing only to prioritize process data transmissions without providing guarantees. None of the systems provide a coordinated medium access and the desired temporal requirements for an NCS as identified in Sect. 1.1.

2.3.6 Existing Limitations of IEEE 802.11

Even though mechanisms to provide QoS guarantees have been added to the IEEE 802.11 standard, recent literature outlined some limitations of the 802.11 QoS mechanisms when different kinds of traffic are supported on the same channel and the total offered workload is high. Some approaches used simulation-based assessments, other ones analytical considerations.

In [122], Moraes et al. showed through simulations that the default parameter values of the EDCA mode are not able to guarantee industrial communication timing requirements, when the highest priority class (AC_VO) is used to support real-time traffic in shared medium environments, where other types of traffic are present. The paper concludes stating that new communication approaches must be devised in order to adopt IEEE 802.11e networks on the factory floor. In [176] it was shown by simulation that it is beneficial to adapt CW_{min} and CW_{max} for the AC_VO class to allow for a larger spectrum of backoff values, thus reducing the number of collisions inside that class. The reason for these results is that the CW_{min} and CW_{max} settings provided by the standard for the AC_VO class determine a narrow range of backoff values for the packets in the class.

Among the works addressing the sensitivity of IEEE 802.11e performance to changes in the CWs depending on the network load there is the one in [184] that, extending the approach in [8], models the EDCA protocol by means of 3-dimensional Markov chains and analyzes the network performance with a varying CW size. Among the approaches proposed in the literature to tune the contention windows there are the adaptive EDCF (AEDCF) proposed in [128], the adaptive EDCA (AEDCA) proposed in [146] and the adaptive technique presented by Vitturi et al. in [177]. While the AEDCA and AEDCF approaches do not provide for changing the values of CW_{min} and CW_{max}, but simply choose the best one in a suitable calculated range, the work [177] proposes an adaptive technique to increase the channel access probability of the highest priority AC in a general industrial scenario, using a fuzzy controller to dynamically find the most appropriate CW range, on the basis of the observed network conditions.

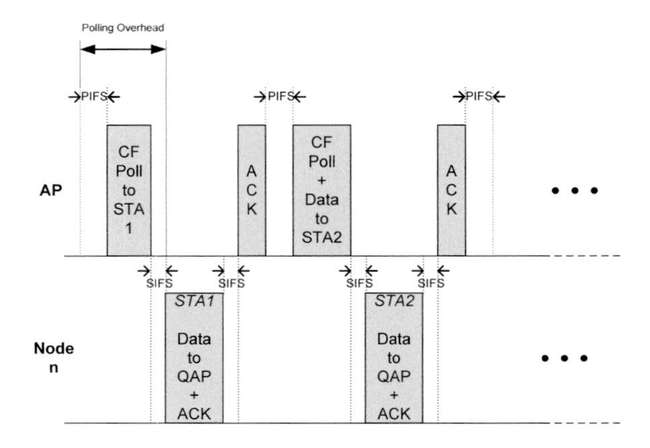

Fig. 2.5: HCCA polling overhead

According to the results in [114] the 802.11e mechanisms are capable of supporting different QoS levels for a wide variety of multimedia applications. Furthermore, the results of [89] have shown that the EDCA might be capable of supporting industrial traffic of real-time class 1 depending on the current traffic load. However, as soon as data communication of real-time class 2 is required, another mechanism such as the polling-based HCCA has to be used [162]. However, polling will cause additional protocol overhead [158] as shown in Fig. 2.5.

Especially for frames with a small payload size, as commonly used on the field level, the HCCA mechanism is rather inefficient, i. e., the protocol overhead is huge as compared to the overall frame duration. The IEEE 802.11 protocol was initially specified with the design goal of providing connectivity for portable and mobile nodes exchanging larger amounts of data. As soon as frame sizes are reduced and frames encapsulate only very small payloads, the protocol overhead drastically increases. The overhead can be calculated as ratio of the transmission time for the T_{OH} and the overall time for the frame exchange sequence T_{seq} according to [70]. The time T_{seq} comprises the data frame, the acknowledgement frame and all interframe spaces. The overhead is shown in Table 2.3 for three characteristical MAC Service Data Unit (MSDU) sizes of a small frame with 20 bytes payload, a mid-sized frame with 200 bytes and maximum frame of 2304 bytes. In the first case more than 89 % of the total transmission duration is consumed by the protocol overhead which wastes a lot of bandwidth and is not acceptable for industrial field level communication with small payload frames.

Table 2.3: Relation overhead to frame sequence length for HCCA (12 Mbps)

Payload Size l_{MSDU} (Byte)	**Payload** $t_{payload}$	**Overhead** T_{OH}	**Frame sequence** T_{seq}	**Overhead** T_{OH}/T_{seq}
20	13.3 µs	112.7 µs	126 µs	89.4 %
200	133.3 µs	112.7 µs	246 µs	45.8 %
2304	1541.3 µs	112.7 µs	1654 µs	6.8 %

This was also realized by the 802.11 working groups and frame aggregation was introduced. Frame aggregation encapsulates more than one data frame destined to the same station with the main objective of increasing the throughput. The performance of the mechanism is evaluated in [100, 97] by analytical and simulation models. Although the throughput increases under ideal conditions, the latency suffers due to waiting times at the aggregating node.

2.4 Wireless Network Topologies

In general, two different topologies of industrial wireless networks exist, wireless infrastructure networks and Wireless Mesh Networks (WMNs). Both topologies are considered in this section and discussed with respect to the application scenario of this work.

Wireless infrastructure networks extend the wired communication network by providing a wireless connection to mobile nodes via APs. In terms of network topology, infrastructure networks can be considered as a simple star topology. Because only point to point wireless connections exist in such a topology, they have major advantages with respect to their communication latency. The main disadvantage of infrastructure networks arise when deploying such networks in large scale installations, e. g., typical for process automation. The necessary wired infrastructure to connect the APs cause tremendous costs and it also results in a very static deployment, due to its inflexibility to deal with structural changes within the plant.

WMNs, on the other hand, are far more flexible in nature and can be considered as a wireless backbone network. They are established by several mesh routers and mesh clients, each of which might have several links forming a mesh topology. WMNs support several self-x features. For instance, their self-organization automatically establishes the wireless backbone without any manual intervention. Their self-healing is able to select alternative routes whenever node or link failures occur. Due to their dynamic characteristics and self-x properties, WMNs are an interesting approach for industrial communication systems [129, 107]. However, the application requirements must be carefully considered before deployment.

Despite of the discussed advantages of WMN in terms of flexibility for the application and large scale deployments, WMNs are not able to satisfy the temporal requirements of the presented application scenario in terms of timeliness and simultaneity. In order to be able to meet those tight temporal requirements, the system must provide an end-to-end communication latency of $\leq 10\,\text{ms}$. Currently, it is possible to achieve round trip times with WMN bounded to approx. 200 ms [106]. Hence, ordinary point-to-point wireless connections as established in infrastructure networks are more promising to satisfy the latency requirements imposed by the application scenario.

Hence, in this work only infrastructure networks are considered. The network consist of several point to point connections between the wireless nodes and a central AP being responsible for the whole coordination of the wireless network.

2.5 Analysis and Classification

A classification of the previously discussed technologies and research activities is shown in Table 2.4. The related work has been classified with respect to their support for the important characteristics real-time class (cf. Sect. 1.1), throughput capacity of the system, capabilities for system integration, the establishment of a global time base, and whether it allows to integrate standard nodes. The following categorization is done for all previously described solutions. If a certain solution fully supports the characteristic, it is indicated with a $\oplus$. If it has only partial support, a $\odot$ is used. Whenever a solution does not support the characteristic, but can be extended the $\ominus$ is used, and in case the solution has no support and is not extendible, the $\oslash$ is assigned.

Table 2.4: Analysis of existing solutions

Solution	Real-time class	Throughput capacity	System integration	Global time base	Standard nodes
IEEE 802.15.4	⊖	⊘	⊙	⊖	⊕
Wireless HART	⊖	⊘	⊕	⊖	⊕
ISA100.11a	⊖	⊘	⊕	⊖	⊕
TSCH/6TiSCH	⊖	⊘	⊙	⊖	⊙
IEEE 802.15.1	⊕	⊘	⊙	⊖	⊕
WISA	⊕	⊘	⊕	⊖	⊘
WSAN-FA	⊕	⊘	⊕	⊖	⊘
iWLAN	⊘	⊙	⊕	⊘	⊘
IEEE 802.11	⊙	⊕	⊕	⊙	⊕

Degree of support for the application scenarios:
⊕ Full ⊙ Partially ⊖ No (extendible) ⊘ No (not extendible)

It can be seen that none of the presented solutions is able to satisfy all requirements. Industrial solutions based on IEEE 802.15.4 are mainly targeting applications of the PA domain, because during the design of IEEE 802.15.4 the focus was put on a very low power consumption of the nodes rather than on real-time capabilities. This is not true for the solutions based on IEEE 802.15.1, because they are able to support real-time class 2. The design and underlying technologies of both systems, 802.15.1 and 802.15.4, and its derivatives do neither provide the basis for a sufficient capacity nor they allow modifications to increase the capacity. Furthermore, the synchronous integration into existing wired RTEs is also a challenge for all solutions except of 802.11, mainly because of the required global time base.

Based on the presented analysis, it is decided to select IEEE 802.11 as the underlying technology for this work. IEEE 802.11 is a suitable candidate, because it offers the potential for integration and the needed capacity. Since it is currently lacking a sufficient support for QoS guarantees, it must be modified in terms of guaranteeing real-time communication services and a global time base. The results of several studies [181, 182, 35] support this decision, because they have also identified this technology to be a promising candidate for future wireless communication systems in industrial automation.

3

Industrial Environment Characterisation

The industrial environment significantly differs from well-known office and home environment. Besides posing several challenges on technologies and systems due to harsh environmental conditions, such as increased temperatures or vibrations, it is remarkably different in terms of deploying wireless technologies. Mainly with respect to two aspects, the industrial traffic characteristics and the industrial wireless channel. Hence, both aspects are further investigated in this chapter by means of extensive experiments in real environments. In Sect. 3.1, two factory automation applications representing the RMS application category are analysed with respect to their temporal requirements. The industrial wireless channel is characterised for both application categories in Sect. 3.2. The results of the channel characterisation are used as input for a realistic model of the wireless channel. The model is presented in Sect. 8.3.2 and used for the evaluation of the proposed solution approach.

3.1 Industrial Traffic Characteristics and Requirements

As already outlined in Sect. 1.1, factory automation applications demand stringent temporal requirements from the communication system, especially on the field level. Replacing an existing wired communication system, either to some degree or completely, by a wireless communication system, requires therefore a detailed analysis of the application and its temporal requirements. An analysis can be done by capturing the communication traffic of the running system. Then, the traffic characteristics are investigated off-line with respect to specific criteria.

In contrast to the IT and office domain, where several activities in this field can be observed (cf. [13]), research about traffic characteristics of factory automation applications is very limited. In [79] a factory communication system was analysed to develop a representative traffic model to assess the performance of the system by means of simulative and analytic processes. The results were based on the distribution of arrival times, distribution of packet length and time characteristics of the outgoing messages. This approach was adapted in this work to determine the requirements of two different manufacturing systems, a manufacturing system for terminal blocks [83] and a robot cell [47]. Both of them include modules which are frequently found in the first application category defined in Sect. 1.1.

The investigated systems deployed heterogeneous communication systems, including Interbus [54], Sercos II [57], Modbus/TCP [58], and standard Ethernet. The update times and latencies of different sensor- and actuator signals are used to derive application characteristics. Subsequently, temporal requirements of both plants are determined using this knowledge. In order to distinguish between multimedia services and traffic of the industrial domain, first multimedia traffic and its requirements is briefly discussed. Afterwards the analysis of both manufacturing systems is presented.

3.1.1 Quality-of-Service Requirements for Common Multimedia Services

The characterisation of multimedia applications and their corresponding traffic is rather advanced, due to the Caida activities [13] and the ongoing standardization (IEEE, IETF, etc.), which is mainly targeted towards IT and office environments. In order to differentiate between the industrial and office domain, the most interesting multimedia applications and their QoS requirements are listed in Table 3.1. A sufficient QoS for multimedia services is mainly assessed by the users perception and might result in accepting or rejecting a specific application. In industrial applications, the process itself determines the requirements, if they cannot be fulfilled, damaging the system or a poor quality of the manufactured product might be the result.

Table 3.1: Performance targets for multimedia applications according to [7]

Application	**Data rate**	**Latency (typ.)**	**Jitter**	**Packet loss rate (PLR)**
Audio (Conversation)	16 ... 64 kbps	$\leq$ 250 ms	$\approx$ 10 ms	10^{2}
Video (uncompressed)	10 ... 100 Mbps	$\leq$ 250 ms	$\approx$ 10 ms	10^{2}
Video (compressed)	64 kbps ... 100 Mbps	$\leq$ 250 ms	$\approx$ 1 ms	10^{11}

The temporal requirements of multimedia applications, such as voice and video transmission, are rather relaxed as compared to the industrial domain, whereas the bandwidth consumption is higher. Especially, video streaming is different; it requires to transmit variable amounts of picture information, which results in variable bit rate (VBR) traffic. However, the traffic pattern can also be considered to have a cyclic behaviour [46].

3.1.2 Analysis and Measurement Methodology

In order to determine communication characteristics and requirements, all communication within a specific system has to be monitored. This can be done at the central PLC, because it is usually connected to all communication systems. The data collection was focussed on the lower two levels, the cell/control level and field level encompassing Interbus [54], Sercos [57], and Modbus/TCP [58]. Interbus is a well-known and widely accepted field bus standard, optimized for the sensor- and actuator area. The Interbus protocol is based on a summation frame mechanism, and it uses a ring topology, i. e., all participating devices are actively involved in the communication. The Sercos II technology has been specifically tailored for motion control, and works with a master-slave principle. The topology is also based on a ring structure being capable of having a single master. Accessing the medium is realized by means of a time slotted approach. As opposed to the other technologies, Sercos uses fiber optics as physical medium. Modbus/TCP was one of the first fieldbuses based on Ethernet and is mainly deployed in factory automation. The protocol is a further development of the Modbus standard, which was initially developed in 1979. The existing Modbus services and models were taken from the old standard and have just been mapped to TCP/IP.

All existing traffic flows of the different protocols must be recorded concurrently in a PC with specific protocol analysis hardware, that has to be clock synchronized. Since real industrial manufacturing processes are investigated, it is not permitted to influence the application during data collection.Moreover, the measurements are conducted passively following a monitoring approach [75] with specific protocol analysis hard- and software. The Sercos Monitor for Sercos II[1], the netAnalyzer for Modbus/TCP[1] and a specific Interbus Analyzer for Interbus.

Afterwards, the captured data is linked with the engineered input- and output-signals of the machine. Thus the temporal behaviour of the metric update time T_{IAT} of individual sensor and actuator signals can be extracted, statistically analysed and assessed regarding their distribution. For every analysed signal the statistical parameters of the update times are determined. These are the mean $\bar{T}_{IAT}$, the median $\tilde{T}_{IAT}$, the mode $\dot{T}_{IAT}$, the minimum value $\min(T_{IAT})$, and the maximum value $\max(T_{IAT})$ with a confidence of 95%. Besides evaluating update times of various process signals, influences of the heterogeneous network structure can be investigated. Finally, the results are classified into the three real-time classes defined in Table 1.1.

3.1.3 Manufacturing System for Terminal Blocks

The first manufacturing system is a typical manufacturing process and is usually used for testing of new technologies and machine concepts. Its bus- and network configuration is shown in Fig. 3.1. The design is quite complex consisting of several PLCs, drives, as well as sensors and actuators. Basically, the system under test can be structured with respect to three involved field bus systems previously mentioned.

[1] `http://de.hilscher.com`

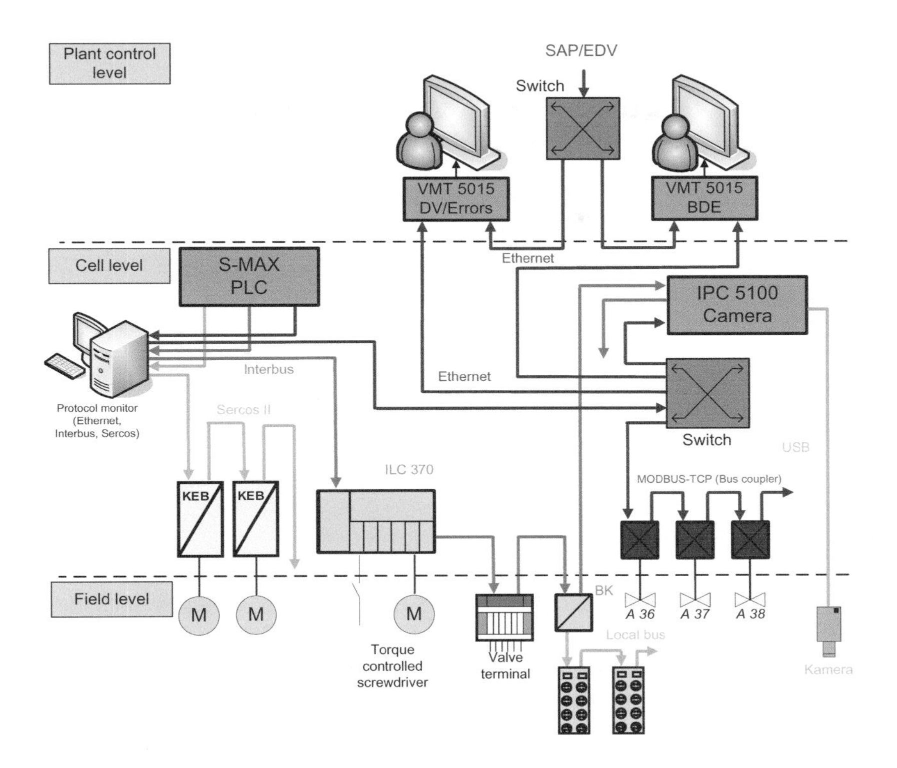

Fig. 3.1: Typical industrial manufacturing system

Almost all drives are controlled by Sercos II, whereas the sensor- and actuator communication is realized with Interbus. Modbus/TCP is used for exchanging parameter and error messages between different subsystems. The overall system is driven by a central PLC based on an industrial PC. Besides all common PLC functionalities, it is equipped with an Interbus and Sercos master. Modbus/TCP utilizes the standard Ethernet interface. The Sercos II master controls two drives, and the Interbus master is connected to a small PLC for driving a torque-controlled screwdriver, several IO devices and another industrial PC. Modbus/TCP is mainly used for monitoring and some safety relevant functionalities of the system.

Technical Process

Since the investigated machine is used for testing and demonstration purposes, some mechanical functionalities are set to be inactive. However, this has no impact at all on the measured communication and traffic characteristics. The overall manufacturing process is divided into several stations. Altogether nine stations are traversed, before the workpiece, which is a connector for printed circuit boards (PCB), is finally assembled and tested. The whole manufacturing process (module 1 to module 9) takes ≈ 165 s. The different stations are supplied with workpieces by a specific workpiece carrier, also referred to as nest. The fine positioning at each station is realized with a servo motor in conjunction with a ball-type linear drive. The return transport is done with a conveyor belt.

The connector cases are fed into the workpiece carrier at module 1. For this purpose a parts catcher takes an empty case out of the supply and inserts it into the carrier. Afterwards the correct position within the carrier is inspected at module 2 using a photo eye. The third station is in charge of putting the wiring pins in the case. Module 4 assures that the wiring pins are properly locked in position by denting them with a pneumatic cylinder. The press out strength is checked at modules 5 to 7, in order to make sure that no wiring pin is released whenever the connector is unplugged from the PCB. All other qualitative features are checked at module 8 with an industrial camera and the IPC 5100 (Fig. 3.1). The last station accepts or rejects the finished work piece. Therefore a parts catcher is used, almost similar to module 1.

Results and Requirements

The first scenario was limited to Interbus and Modbus/TCP. The overall measurement duration was 2 hours and determined a priori by exemplary measurements (cf. [75]). The main objective was to investigate individual sensor and actuator signals with respect to their update times and resulting response times. All three field bus systems were considered in the second scenario. The measurement duration was set to $7\,min$, due to the maximal possible time of the Sercos Monitor, caused by its limited ring buffer size. It was carried out with having the goal to show signal dependencies in heterogeneous communication networks and resulting latencies.

The sensor and actuator signals have been extracted from the collected data beginning with the first scenario. The distribution of the update times is shown as histogram $H(T_{IAT})$ with bin size k and sample size N, as well as empirical Cumulative Distribution Function (CDF) $F(T_{IAT})$.

The signal *ST1_OUT_xOeffner_2_vor-Y* (Interbus) controls the movement of an actuator of module 1. It belongs to category *high requirements*.

The bimodal distribution as shown in Fig. 3.2 (a) can be explained with the sequential operation of the manufacturing process. It can be distinguished between active and passive phases. In the active phase the workpiece is processed and in the passive phase no processing takes place. In Fig. 3.2 (b) the histogram for values of ≤ 50 ms is shown, i. e., the active phase. The histogram shows an increased number of values for 23 ms and 27 ms.

The signal with the highest update time within the manufacturing system was identified at station 3. It is the sensor signal *ST3_IN_xStift_vorhanden-B* (Interbus) coming from a photo eye. It is used for quality assurance, i. e., to check whether a wiring pin has been properly put into the case or not. The histogram for all values ≤ 10 ms is depicted in Fig. 3.3 (b). An increased number of values can be identified at 3.37 ms and 6.5 ms. Especially the peak at 3.37 ms should be noticed, since this value corresponds to the Interbus cycle time. This leads to the conclusion, that the signal reaches the maximum possible update time of the field bus.

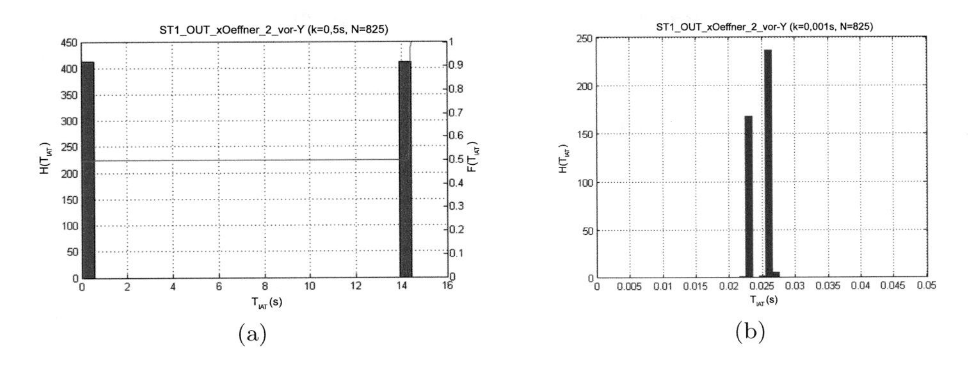

(a) (b)

Fig. 3.2: Update time distribution of the signal *ST1_ OUT_ xOeffner_ 2_ vor-Y*

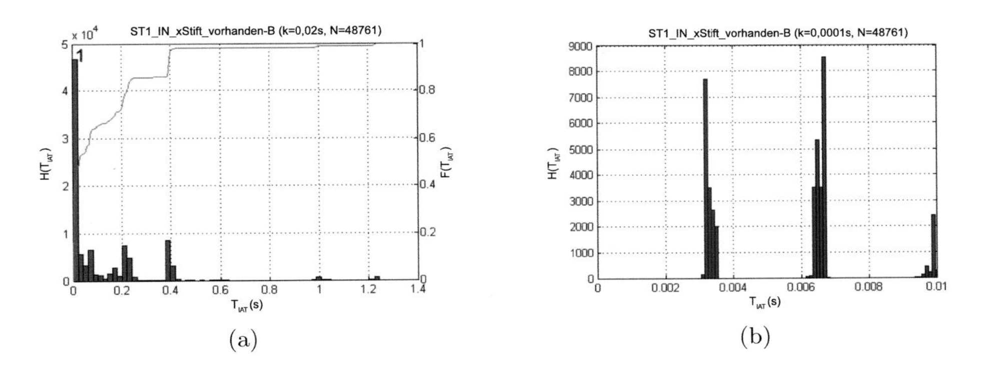

(a) (b)

Fig. 3.3: Update time distribution of the signal *ST3_ IN_ xStift_ vorhanden-B*

However, this seems to be sufficient, since no errors are observed during operation of the system.

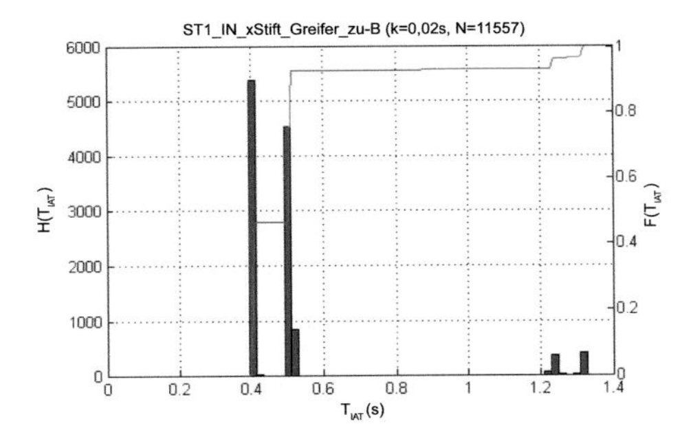

Fig. 3.4: Update time distribution of the signal *ST3_ IN_ xStift_ Greifer_ zu-B*

Signal *ST3_IN_xStift_Greifer_zu-B* has been investigated to serve as an example for the category *medium requirements*. The corresponding update times are shown in Fig. 3.4. Peaks of the measured values can be observed for 400 ms and 500 ms. The difference between both peaks is evidently smaller as compared to the signals above, which can be explained by the almost similar duration of active and passive phases of the manufacturing process. A summary of the three analysed signals is provided in Table 3.2 in terms of their statistical parameters.

Table 3.2: Summary of the update times of different analysed signals

Signal	**Sample size N**	$\bar{T}_{IAT}$ (s)	$\tilde{T}_{IAT}$ (s)	$\dot{T}_{IAT}$ (s)	**min**(T_{IAT}) (s)	**max**(T_{IAT}) (s)
ST1_OUT_xOeffner_2_vor-Y (A)	825	7.17	0.0266	0.023	0.0225	14.44
ST3_IN_xStift_vorhanden-B (S)	48761	0.12	0.0263	0.0039	0.0039	1.24
ST3_IN_xStift_Greifer_zu-B (S)	11557	0.51	0.499	0.405	0.39	1.32

Besides the update time T_{IAT}, mainly characterising the overall process itself, the response time T_{IO} is of major importance for deriving temporal requirements (cf. Sect. 4.3). It is defined as the time when the transmission has been started at the sensor source (S) until it is received at its actuator destination (A). Selected sensor and actuator relationships are shown in Table 3.3. Requirements belonging to real-time class 2 are found in modules 3 and 4 (in the range of 10 ms), whereas low requirements of 270 ms belonging to real-time class 1 are explored for module 9.

Table 3.3: Response times of different analysed signals

Source	**Destination**	T_{IO} (s)
ST1_IN_xOeffner_2 (S)	*ST1_OUT_xOeffner_2 (A)*	0.016
ST1_IN_xGeh_greifen_zu (S)	*ST1_OUT_xGeh_greifen_zu (A)*	0.042
ST3_IN_xStift_Greifer_zu (S)	*ST3_OUT_xStift_Greifer_zu (A)*	0.023
ST3_IN_xVereinzelung_unten (S)	*ST3_OUT_xVereinzelung_ab (A)*	0.01
ST4_IN_xEindruecker_vorne (S)	*ST4_OUT_xEindruecker_vor (A)*	0.01
ST9_IN_xEntnahme_vorne (S)	*ST9_OUT_xEntnahme_vor (A)*	0.27

The mutual dependence of the signals *xDUSwivelUnitInFront-B* and *xDU1SwivelUnitBack-B* (both connected to Interbus) versus the signal of the main drive (Sercos II) is shown in Fig. 3.5, in order to show response times when several field buses are utilized.

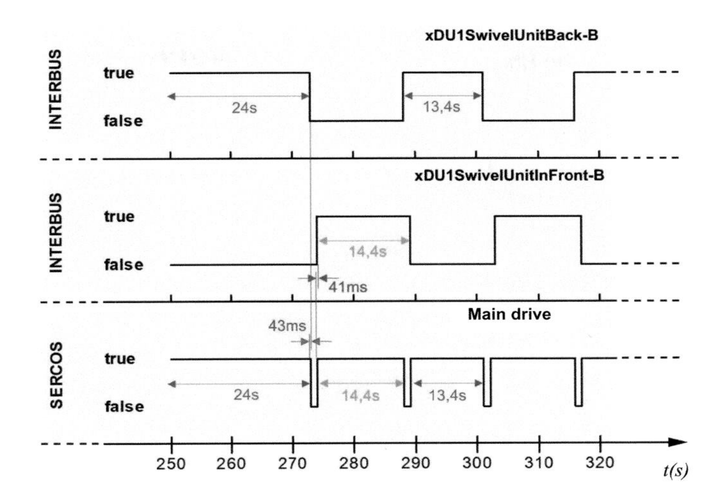

Fig. 3.5: Response times for the heterogeneous structure

Both Interbus signals are for the cylinder switch of the swivel units rotary drive. The signals are alternately *true* depending on the direction of motion. From the time dependent course of the main drive signal it can be inferred, that the long time intervals (signal *true*) were due to the transportation of the workpiece carrier and the short ones (signal *false*) were caused by changing it. For the main drive and sensor signal *xDU1SwivelUnitInFront-B* a mean response time of 45 ms was determined. The sensor signal *xDU1SwivelUnitBack-B* and the main drive revealed a mean response time of 48 ms which belongs to real-time class 1.

3.1.4 Robot Cell

The second analysed system is a robot cell performing a pick-and-place task and is less complex. The robot cell is usually used for demonstration and testing purposes. It is shown in Fig. 3.6(a).

Technical Process

The robot cell consists of a robot arm (Kuka 125/2-ZF) and two conveyor belts. The two conveyor belts transport the work piece in opposite directions. As soon as a work piece reaches the end position of a conveyor belt, the robot picks up the work piece and places it on the other belt. Hence, the work piece movement can be considered as circular.

The second system consists of three PLCs which are interconnected by Interbus and Ethernet [67]. The main PLC is responsible for the robot arm and the mounted robot tool. The communication with the robot is realised by an Interbus system coupler for PC–systems connected to the main PLC. The second and third PLC

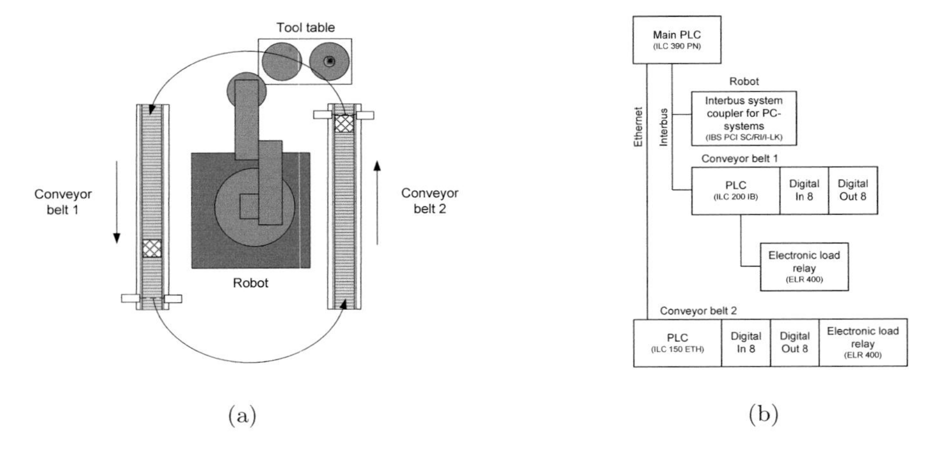

(a) (b)

Fig. 3.6: Pick-and-place robot application and its automation architecture

are used for the first and second conveyor belt. The second PLC communicates with the main PLC via Interbus, the third PLC communicates with the main PLC via an ordinary Ethernet connection (cf. Fig. 3.6(b)).

Results and Requirements

In the second system the signals of conveyor belt 1 and the signals of the PLC controlling the robot are analysed. It can be assumed that conveyor belt 1 and 2 perform the same task. The PLC for the conveyor belt has 5 process signals, i) *Conveyor belt start occupied (input)*, ii) *Conveyor belt end occupied (input)*, iii) *Conveyor belt start–region occupied (input)*, iv) *Conveyor belt end–region occupied (input)*, and v) *Conveyor belt run (output)*. The robot is controlled by parameterised command words. The performed operation of the robot depends on the bits set in the command word and the bits set in the two parameter words. The analysis showed only two different commands; one command was responsible for positioning tasks, and the other command was concerned with opening and closing tasks of the robot tool. The following signals (bits) have been observed, because all other signals did not change:

- Command–word: bits 0 and 1
- Parameter–word 1: bits 0 to 2
- Parameter–word 2: bit 0

The results of the analysis of conveyor belt 1 revealed a minimum signal duration of 0.9184 s for the input signal *conveyor belt end–region*. The update times (T_{IAT}) of all input signals have a value of approximately 27.8 s and are the signature of this process which indicate a cyclic behaviour. Because the signal durations and update times for the conveyor belt are within a scale of seconds, they can be classified as signal of real-time class 1.

The second analysis is focused on two signal dependencies in the system, both responsible for stopping the conveyor belt. In the first case, the conveyor belts can be stopped by their PLCs, if the work piece reaches the end position with the signal *Conveyor belt end occupied.* The minimum value of this response time is 25.1 ms. In the second case, the PLC which controls the robot stops the conveyor belt, if the robot is going to pick up the work piece. The minimum response time in this case, i. e., after resetting the *Conveyor belt 1 enabling signal* by the robot PLC, was 8.6 ms. The maximum value of the response time is determined to be 18.1 ms. By considering all values of the response time, it is observed that most of the values are $\geq$ 10 ms. Therefore, this process belongs to the category real-time class 1.

3.2 Industrial Wireless Channel

In this section the characterisation of the real industrial channel is presented. The channel characterisation is focused on the relevant environments for the both application categories. The channel of category 1 reconfigurable manufacturing system is represented by a typical manufacturing environment including a robot cell. The wireless channel characterisation for category 2 is represented by a wind energy plant. The results of the characterisation are important for two reasons, (i) the system design of the isochronous wireless network, and (ii) the design of a realistic model of the wireless channel to increase the credibility of simulation case studies for wireless systems.

A common characteristic of a wireless channel is the frequency response, i. e., the frequency-dependent transfer function. It expresses the attenuation and the phase of a transmitted sinusoidal signal for a given spectrum with bandwidth B. One of the main differences of an industrial wireless channel, in comparison to office environments, are the increased multi-path effects caused by metal elements, heavy machinery and several moving objects. Multi-path propagation results in a time and frequency variant behaviour of the channel transfer function, and have an impact on the reliability of the communication in terms of the Bit Error Rate (BER), the Packet Error Rate (PER). If no measures are introduced by the wireless communication system, both are resulting in an increased PLR perceived by the application.

The most common approach to measure frequency responses are realized by a Vector Network Analyzer (VNA), which uses a spectrum analyzer together with a swept sinusoidal source working in a synchronized tracking mode. The approach suffers mainly from two weaknesses. The first issue is the lack of mobility. To measure the frequency response for a given spectrum, the receiver needs to know the transmitted signal. This is done by synchronizing the transmitted and the received signal using a direct wired connection. However, with rotating and moving objects, such as the wind energy plant application, as well as in many other industrial application environments a wired connection for synchronization is impossible. The second issue is the lack of real-time characterisation. To measure the whole spectrum, the approach sweeps the frequency of the sinusoidal signal from the lower boundary of the spectrum to the upper boundary. Thus, the whole spectrum is not measured simultaneously and therefore it does not characterize the wireless channel in real-time.

3.2.1 Weaknesses of Existing Channel Characterisation Approaches

The existing approaches to characterise a wireless channel can be divided into (i) narrowband and (ii) wideband sounding [27, 136, 142].

For narrowband sounding mainly the single tone technique is applied. The single tone technique comprises a transmitter and a receiver. The transmitter excites an unmodulated RF carrier, while the receiver measures the amplitude and phase variation of the received carrier. Therefore the frequency and the phase reference of the receiver must be synchronized with the transmitter's reference. The synchronization is normally realized with a cable for direct connection, such as in a VNA. The simplicity, the accuracy and the inexpensive narrowband equipment are the advantages of this well-established technique, but it suffers from two limitations: It has (i) no opportunity for real-time measurements and (ii) mobility is not provided. For this work, the second limitation of mobility is more important. It stems from the required direct cable connection between transmitter and receiver. To overcome the limitation of mobility, the single tone technique can be extended by wireless synchronization as proposed by Molina et al. in [121]. The work proposes an implementation using a GPS-based reference clock synchronization with two VNAs. However, such an extension suffers from its complexity and its expensive synchronization equipment.

In contrast to narrowband sounding, wideband sounding approaches enable time-variant wireless channel characterisation. They can be divided into (i) periodic pulse techniques and (ii) pulse compression techniques [136]. The periodic pulse technique transmits a periodic pulse train. After reception, the variation of the frequency representation can be calculated. To characterize frequency-variant wireless channels with high accuracy, the period duration must be above all multipath delays. Further, to assume a non-varying wireless channel, the period duration shall be much below the channel's variation. For example, a wireless channel which is periodically disturbed by a moving obstacle each 100 ms can be successfully captured with a pulse repetition of 1 ms. To determine the reception delay, the transmitter synchronizes the reference clock with the receiver directly via cable. The technique benefits from wideband measurement features, which enable real-time characterisation of wireless channels. However, the limitation of mobility is present, because the synchronization uses a direct cable connection.

Another option is to reduce complexity by removing the synchronization concept. Then, the post-processing returns only the magnitude of the frequency response, which disables the determination of the impulse response. This simplification permits mobility with inexpensive equipment and is the preferred approach in this work.

3.2.2 Approach to Wireless Channel Characterisation

Since all techniques presented above suffer from mobility limitations due to the required wired connection for synchronization, an approach based on the periodic pulse technique without synchronization is presented in [9].

Principle

The discussed approach uses a signal waveform following two properties [133]: First, a signal which is discrete in the frequency domain is periodic in the time domain. Second, the duration of an impulse is inversely proportional to its bandwidth in the frequency domain. These two principles result in a periodic pulse signal.

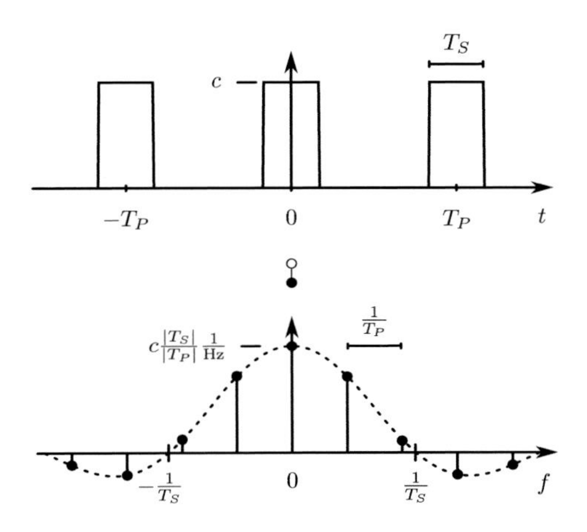

Fig. 3.7: Periodic rectangular signal in time domain and frequency domain

Figure 3.7 illustrates the principle with a rectangular pulse waveform as baseband signal. It shows a periodic pulse signal in the time (upper diagram) and frequency domain (lower diagram). Three conclusions can be drawn:

- The period duration T_P characterizes the discrete frequency spacing with $\frac{1}{T_P}$.
- The inverse of the pulse duration T_S characterizes the bandwidth of the envelope to such an extent that its zero crossings are multiples of $\frac{1}{T_S}$ and the maximum bandwidth of the envelope – or the main lobe – is $\frac{2}{T_S}$.
- The ratio between the (im-)pulse duration and the period duration $\frac{T_S}{T_P}$ characterizes the proportional envelope height in the frequency domain.

This leads to the following tendencies:

- To have minimum discrete frequency spacing, the period duration shall be as large as possible

$$T_P \to \infty.$$

- To have the maximum bandwidth in the frequency domain, the pulse duration shall be as small as possible
$$T_S \rightarrow 0.$$
- To have the maximum height of the frequency domain envelope, the pulse duration shall be as large as possible.

Based on the principle and its tendencies a transmitter is designed and described in the following section.

Measurement Setup

The whole measurement setup consists of three parts: (i) transmitter, (ii) receiver and (iii) post-processing unit (see Fig. 3.8).

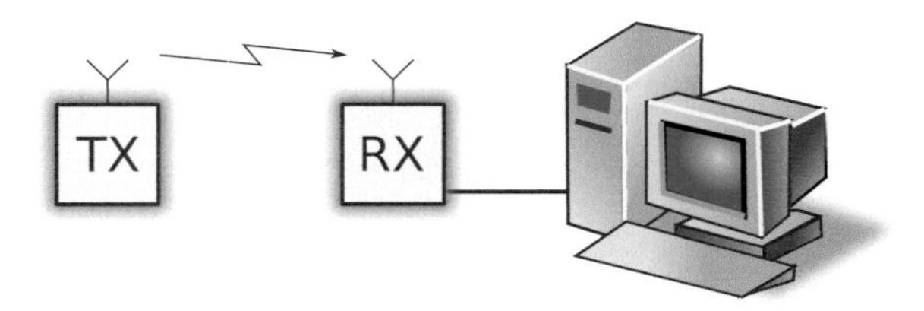

Fig. 3.8: Measurement setup with transmitter (TX), receiver (RX) and post-processing (computer)

The transmitter is a mobile device which can be mounted on rotating, driving or otherwise moving objects. It is a software defined radio (SDR), configured to run in a standalone mode instead of being controlled by an external computer. The mobile transmitter is able to perform various modulation techniques. To satisfy the prerequisites of the modulation techniques, it has a wideband radio frontend, being used to transmit the required wideband signal.

The transmitted signal differs from the ideal waveform shown in Fig. 3.7 both in the time domain and in the frequency domain. To come close to the ideal waveform, all data samples in the buffer are zero except the first one, which has the maximum value. This results in a frequency representation shown in Fig. 3.9, using a buffer size of 64 samples and a center frequency of 2.44 GHz. The transmitted signal is captured by the Real-time Spectrum Analyzer (RSA).

To increase the Signal-to-noise Ratio (SNR), only 40 MHz of the reference signal are used. Hence, the applied frequency bandwidth in Fig. 3.9 ranges from 2.42 GHz to 2.46 GHz. In this case the number of characterising frequencies is reduced to 45, but the SNR increases to approximately 35 dB. This also enables large-distance measurements with deep drops in the frequency response.

The receiver is a RSA Tektronix® RSA6114A. It stores the received samples in real-time and calculates a spectrogram. It works in two operational modes: acquisition and analysis. Compared to the analysis mode, the acquisition mode has a lower

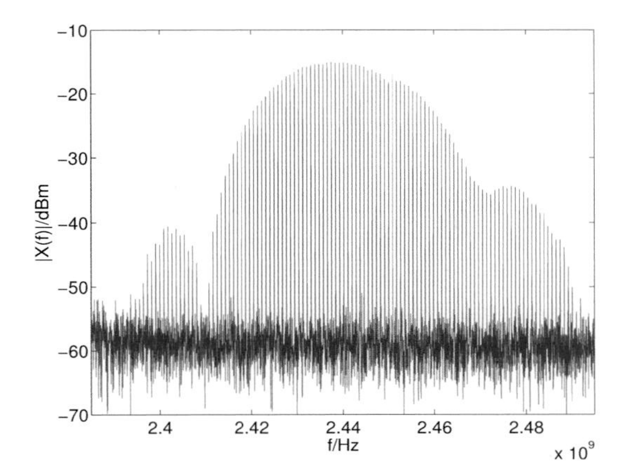

Fig. 3.9: Measured spectrum of the reference signal with 64 samples buffer size, 2.44 GHz center frequency and 893.2 kHz average discrete frequency spacing

minimum time resolution. However, the maximum measurement time results from the analysis mode. Hence, the optimal operation mode depends on the variation speed of the wireless channel and on the required measurement time. Looking at the whole setup, the minimum time resolution is limited by the transmitter which results in approximately 1.12 µs. For instance, using the maximum bandwidth of 47.3 MHz results in the maximum measurement time of 3.58 s (acquisition mode; RSA bandwidth of 60 MHz). For long-term measurements the analysis mode can be chosen with a trade-off in time resolution.

Finally, the spectrogram has to be processed to return the time-dependent frequency response. This is performed offline, i. e., after the measurements have been conducted, on a computer using Matlab from MathWorks®, Inc.

3.2.3 Channel Characterisation

The previously described approach has been applied in two different industrial environments to characterise their wireless channel. Both industrial environments and the obtained results are described in the following subsections. The obtained transfer function will be used in this work as basis for a realistic wireless channel model for simulation case studies.

Manufacturing Environment with Robot Cell

The first environment was a typical factory floor with robot cell, representing the first application category (cf. Sect. 1.1). The mobile transmitter is mounted on a moving robot arm as shown in Fig. 3.10, which repeats a movement on a path marked by four points with a cycle duration of 1.5 s. The robot arm movement has a diameter of approximately 1 m and moves with a speed of approximately

2 m/s. The distance between the receiver and the robot arm is approximately 3 m. To have the largest bandwidth, the measurement is performed using the reference signal shown in Fig. 3.9 with the center frequency of 2.44 GHz and the buffer size of 64 samples. To perform long-term measurements with the relatively slow robot arm, the receiving RSA operates in the analysis mode with 801 trace points. The time resolution is chosen to be 1 ms, which results in a maximum measurement time of 60 s. The actual measurement time was approximately 10 s.

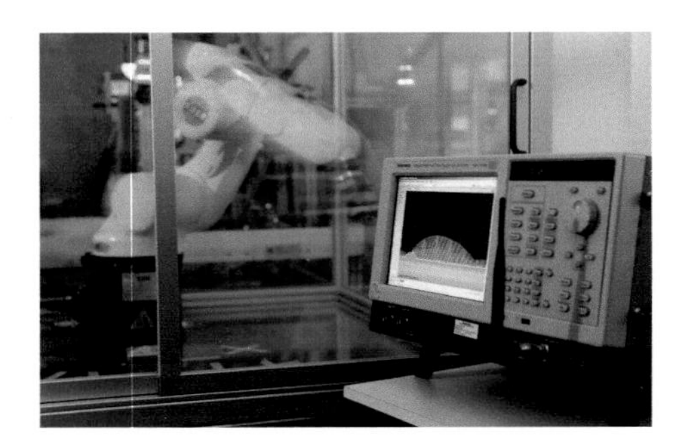

Fig. 3.10: Robot arm (left) with mounted transmitter and the RSA as receiver (right)

The time-variant frequency response is shown in Fig. 3.11. It is derived by comparing the transmitted and the received signal. The horizontal axis represents the frequency from 2.42 GHz to 2.46 GHz. The vertical axis represents the measurement time up to 10 s. The colour represents the logarithmic channel gain according to the colour bar ranging from -40 dB (red) to -65 dB (blue). The maximum variation of the channel gain is 25 dB.

Wind Energy Automation Environment

The second environment is a typical wind energy plant, representing the second application category (cf. Sect. 1.1). The mobile transmitter is mounted within the rotor of the system, the receiver is placed in the static nacelle as depicted in Fig. 3.12(a). The transmitter antenna is mounted on top of an existing control cabinet for the components of the control system. The receiver antenna is situated in the front part of the nacelle. The maximum separation of both antennas during the measurements is 7 m with either Line-of-sight (LOS) or Obstructed Line-of-sight (OLOS) links. The antenna positions are shown in Fig. 3.12(b). The system rotates with a nominal speed of 20 rpm, i. e., one full rotation is completed within 3.5 s.

To have the largest bandwidth of 47.3 MHz, the measurement is performed using the reference signal shown in Fig. 3.9 with the center frequency of 2.437 GHz and the buffer size of 64 samples. The receiving RSA is operated with a bandwidth of

40 MHz and operates in the acquisition mode. The time resolution is chosen to be 1 ms, which results in the actual measurement time of 5.4 s.

The time-variant frequency response of the wind energy plant is shown in Fig. 3.13. It is again derived by comparing the transmitted and received signal. The horizontal axis represents the frequency from 2.42 GHz to 2.46 GHz. The vertical axis represents the measurement time up to 5.4 s. The colour represents the logarithmic channel gain according to the colour bar ranging from -60 dB (red) to -90 dB (blue).

Discussion

The first environment shows a frequency response which is varying in time as well as in frequency. The variation has a span up to approximately 25 dB. Further, the figure shows fast variation especially during the first 6 s and slow variation during the last 4 s, while the robot arm paused. Such a wireless channel is very challenging for the given wireless system and an increased BER and decreased reliability of the transmissions can be expected.

The obtained transfer function for the second environment is mostly varying in time. The frequency variances are very small leading to the conclusion that almost

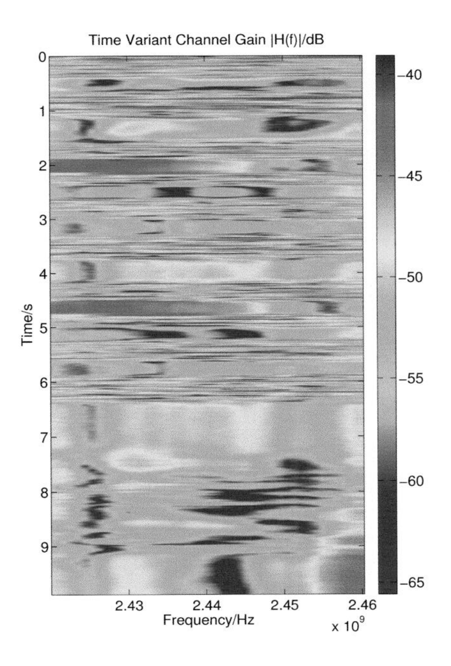

Fig. 3.11: Time and frequency variant wireless channel between a static position and a moving robot arm

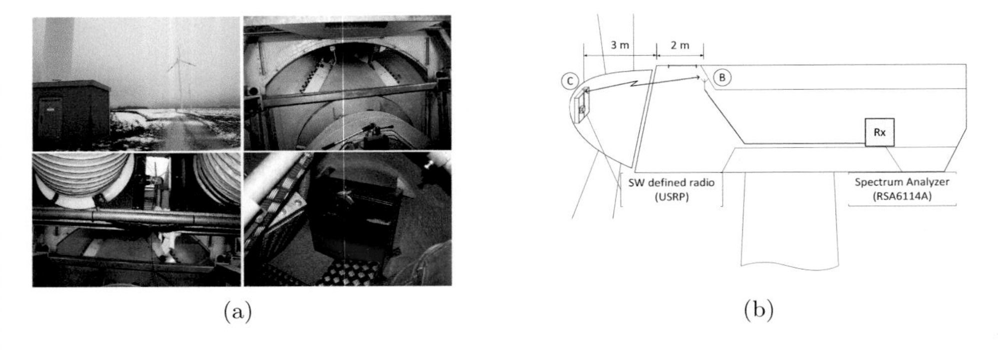

Fig. 3.12: Wind energy plant and measurement setup

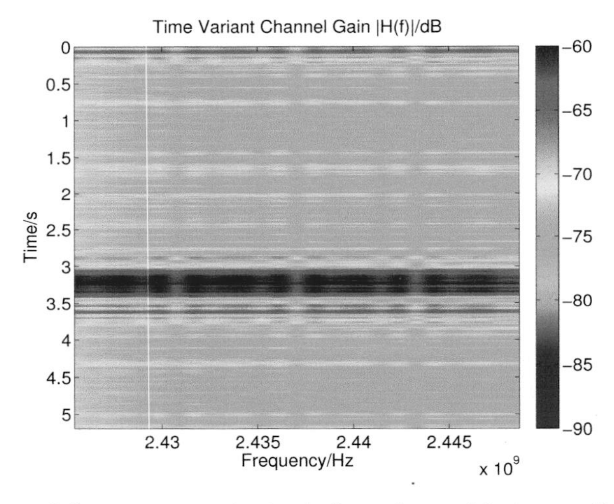

Fig. 3.13: Time and frequency variant wireless channel between the static nacelle and the moving rotor

no multi-path effects are present in this environment. The variations of the transfer function span up to approximately 18 dB and a periodic pattern with a period of approx. 3.5 s is observed. The pattern is caused by the rotation of the system. The maximum values of the channel gain are -60 dB and achieved for an equal orientation of both antennas in horizontal and vertical direction. As soon as the rotation causes a deviation the channel gain decreases to values between -70 dB and -75 dB. Since the frequency dependent behaviour of this channel depends mostly on the position of the rotor, the channel can be considered as flat and has less impact on the wireless communication system [142].

4

Isochronous Wireless Network for Industrial Automation

In this chapter the solution approach of an isochronous wireless network for real-time communication in industrial control applications is presented. The system is based on IEEE 802.11 [70] to allow a seamless integration of the wireless system into existing RTEs and to ensure a sufficient capacity. The approach has been published in [164, 168].

In the next section the system model for the proposed isochronous wireless network is defined. The system model provides the basis for further analysis. It defines the environment and considered constraints and assumptions for the solution approach. It is used to derive relevant metrics for the isochronous wireless network and specifies a fault model. The solution approach is described in Sect. 4.2. The description includes all relevant conceptual components, their interactions as well as the specific problems addressed by them.

4.1 System Model

Before the different components of the solution approach and their interactions are described, the considered system model is presented in this section. First, a network model is introduced in Sect. 4.1.1. It consists of a wireless communication network including specific assumptions for the wireless links, and a wired communication network. The QoS parameter for control applications are described in Sect. 4.1.2. This includes relevant metrics for the application and a traffic model. Finally, a fault model defines the constraints for the system and the assumed failure semantics.

4.1.1 Network Model

The basic network topology is shown in Fig. 4.1. It consists of a hybrid wireless and wired network.

Wireless Network For the IWN a star topology is assumed where the AP is the central node being connected to the wired network and to the central PLC accordingly. It further consists of a set of n wireless nodes $\mathcal{N} = \{N_1, N_2, \ldots, N_n\}$, which include a single IOD. Each node N_i sends and receives a set of real-time traffic flows $F_i = \{f_1, f_2, \ldots, f_n\}$ with the PLC, where $F_i \subseteq \mathcal{F} = \{F_1, F_2, \ldots, F_n\}$ being $\mathcal{F}$ the set of all real-time traffic flows in the wireless system. It is assumed that the propagation delay for each node N_i is approximately the same in both directions.

The central AP is coordinating the medium access on a cyclic basis by using a cyclic structure with a given period called Network Update Time (NUT) T_{NUT}. It is also assumed that all wireless nodes are using the same channel and the data rate for real-time traffic flows is fixed to 12 Mbps to ensure the usage of a very robust modulation type. Due to the importance of the AP as central coordinator of the IWN, it is assumed that the AP is always available. This could be achieved by a physical redundancy concept for the AP, such as duplicated APs in a hot-standby mode. However, such a concept is not part of this work.

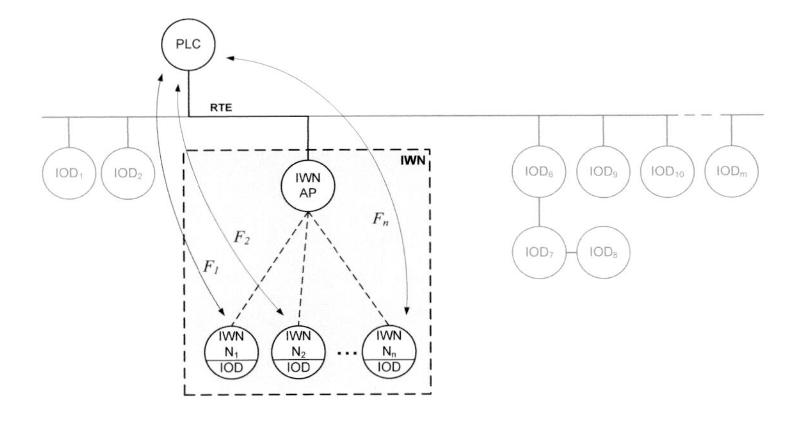

Fig. 4.1: Topology of the network model

Wireless Link The dynamic behaviour of the wireless channel might pose severe constraints on the wireless communication system. Due to the dynamic environment, which frequently changes and is heavily influenced by moving objects, a link might completely fail for a long time or is temporarily not available. Especially in systems with mobile nodes, such as AGVs, the nodes might reach areas of a factory where a sufficient wireless coverage can not be guaranteed any more.

However, there are also systems that can be considered as rather static although they require a wireless communication system. The relevant application categories for this work belong to this kind of applications. They are rather static, i. e., the wind energy plant and modular manufacturing systems encompass only moving nodes, but can not be compared with a mobile system like an AGV. For example, the wind energy plant represents an application where the wireless system is used to replace slip rings and the operational area for the wireless nodes is limited by definition. This is also true for many other applications in factory automation that are targeting slip ring replacements.

Regardless of the previous discussion, the relevant application scenario of this work also encompasses a time and frequency varying channel resulting in an arbitrary number of omission failures (*omission degree OD*) at some node location. As described in the fault model, the system is able to handle omission failures as long as the omission degree is bounded. Otherwise, the system is no longer able to satisfy the temporal requirements. Hence, in this work we follow the approach presented by Schemmer in [154]. The wireless coverage of the system is separated in three different areas, called *valid*, *partially valid*, and *invalid*. The valid area is located around the AP up to a certain distance d_{valid} which varies depending on the current channel conditions. It is assumed that all nodes within the valid area are able to communicate with a bounded omission degree. In the partially valid area, nodes can still communicate, but with an unbounded number of omissions. This will probably result in a vast amount of retransmissions without being able to provide any temporal guarantees. Within the invalid area, nodes can not communicate at all. It is assumed that the valid area has a bounded omission degree of $OD_{max} \leq 2$, which is also tolerated by common RTEs. Due to the aforementioned static nature of the application and the provided results in Sect. 6.2.1, this assumption can be considered as realistic.

Moreover, it is assumed in this work, that there are no co-existence problems within the used frequency spectrum to avoid interferences with other technologies as well as with other IEEE 802.11 networks. This can be achieved by organizational measures such as a frequency planning and frequency management for all used technologies within the environment.

Wired Network Considering the specific mechanisms of some RTE to guarantee real-time communication and its associated cyclic behaviour is not part of this work. Due to the synchronization of the wired and the wireless system, it is assumed that the RTE scheduling is able to provide an inbound or outbound timeslot for communication whenever the uplink or downlink of the wireless network is required at the AP. This must be probably guaranteed and considered during the commissioning phase of the wired network, of course depending on the requirements of the application.

4.1.2 QoS Parameter for Control Applications

An NCS for closed-loop control is considered in this work as application. It consists of a central controller, referred to as PLC, and several distributed sensors and actuators, which are connected via IODs. The sensors sense the current status of the controlled technical process, and provide their values to the controller resembling a consistent image of the process. After the sensor values have been processed within the controller, the updated set points are sent to the actuators. The PLC establishes at least two logical connections to each of its IODs. The IOD receives a traffic flow and acts as consumer for actuator values and at the same time produces sensor values as producer being the second connection.

For closed-loop control, i. e., the defined application scenario, it is important to capture the process image at all nodes $\mathcal{N}$ at a defined point in time called input valid time t_{in} with a defined instantaneous jitter $J(t_{in})$. This guarantees a system wide consistency of process data. The same holds true for providing the updated values to the actuators, all set point values must be provided to the actuators at the output valid time t_{out} with a defined instantaneous jitter $J(t_{out})$ at all nodes $\mathcal{N}$. This is usually referred to as isochronous behaviour and defined as follows.

Definition 4.1 (Isochronous) *All nodes $N_i \in \mathcal{N} : 1 \leq i \leq n, n \in \mathbb{N}$ are defined to be isochronous for the application, if the update time T_{IAT} of their t^i_{in} and their t^i_{out} has a jitter J which is less than a defined constant ξ.*

$$\forall t_{in} \in \mathcal{N} : J(T_{IAT}(t_{in})) \leq \xi, \{\xi \,|\, \xi \in \mathbb{R}\} \tag{4.1}$$

$$\forall t_{out} \in \mathcal{N} : J(T_{IAT}(t_{out})) \leq \xi, \{\xi \,|\, \xi \in \mathbb{R}\} \tag{4.2}$$

In this context, two different properties of the real-time communication system are distinguished, simultaneity and timeliness. Simultaneity defines the allowed deviation of consecutive events in a distributed system. As corresponding QoS parameter the jitter of the update time, as defined in Def. 4.6, is considered.

Definition 4.2 (Simultaneity) *Simultaneity is defined as the execution of concurrent events in a distributed system, such as reading sensor values or writing actuator values, with a defined temporal deviation that is $\leq \xi$, determined by the application.*

Timeliness is defined as the successful transmission of data within a given time bound. The time bound is determined by the technical process and its dynamics. The relevant QoS parameter for timeliness is latency as defined in the next paragraph. Hence, timeliness assures that the received application data is still valid.

Definition 4.3 (Timeliness) *Given a deadline d_i associated to message m_i. The timeliness requirement is defined to be satisfied, if*

$$t_i^{Rx} \leq t_i^{Tx} + d_i, \tag{4.3}$$

where t_i^{Rx} is the time of reception of m_i and t_i^{Tx} is the time when m_i is generated by the application.

Traffic Model In order to model the traffic for this work the notion of real-time traffic flows is introduced. Real-time traffic flows can be of type cyclic and acyclic. In this work the main focus is put on cyclic real-time traffic (cf. Sect. 6.1.2). It is the most frequently used industrial traffic type and usually superordinated to acyclic real-time traffic [42]. Cyclic real-time traffic flows are exchanged between the PLC and IODs. In addition to this, mainly real-time traffic flows with small payload sizes $l \leq 100$ bytes are considered for process data exchange, because they are most important for cyclic real-time traffic [42].

The application within each node $N_i \in \mathcal{N}$ generates and receives a set of cyclic real-time traffic flows $F_i = \{f_1, f_2, \ldots, f_n\}$, where each traffic flow consists of a downlink and an uplink flow $f_i = \{f_i^{DL}, f_i^{UL}\}$ both forming a set of periodic messages $\mathcal{M} = \{m_1, m_2, \ldots, m_n\}$. Each real-time traffic flow f_i is characterized by a QoS parameter set denoted as tuple $Q_i = \langle T_{P_i}, d_i, OD_i^{max}, l_i, J_i \rangle$. It is assumed that Q_i is identical for uplink and downlink direction.

- **Send period** (T_{P_i}): The send period denotes the time interval of generating a new message. Since the considered flows are cyclic, the period is a constant value.
- **Deadline** (d_i): The deadline d_i of some message m_i is the latest time instant, relative to its release time, when the message must be delivered to the receiver. It is assumed in this work that $d_i = T_{P_i}$, i. e., if m_i is released at time t_0, $d_i = t_0 + T_{P_i}$.
- **Omission degree** (OD_i^{max}): The omission degree specifies the maximum number of consecutive omissions that are tolerated by the application.
- **Payload size** (l_i): The payload size denotes the MSDU size in bytes. Considering the additional protocol overhead and the data rate of the wireless system, the complete transmission duration C_i is calculated based on l_i and the protocol overhead for scheduling and admission control.

Based on the real-time traffic flow characterization, different relevant metrics, which are used throughout this work, are derived. A graphical representation is shown in Fig. 4.2 for the downlink direction from the PLC to the IOD. The first introduced metric is the latency T_{Lat}, denoted as T_{Lat}^{DL} for the wireless downlink direction and as T_{Lat}^{UL} for the wireless uplink direction. The overall latency T_{Lat} is defined as follows.

Definition 4.4 (Latency) *The latency T_{Lat} is the time interval from starting an application data transmission of message m_i at the producer until a successful reception of the message at the consumer.*

$$T_{Lat} = t_i^{Rx} - t_i^{Tx} \tag{4.4}$$

The latency can be further divided into T_{Lat}^{DL} or T_{Lat}^{UL} for the wireless part and T_{Lat}^{RTE} for the wired part. As already introduced, a predictable latency is very important for the NCS. The achievable latencies T_{Lat} of both directions have a major influence on the overall response time T_{IO} that is very important for the technical process. This will be discussed in Sect. 4.3.

The second metric is the update time T_{IAT}, either denoted as T_{IAT}^{DL} for the downlink direction or as T_{IAT}^{UL} for the uplink direction. It is especially important for industrial control applications, since they rely on a periodic sampling of the technical process, i. e., cyclic process data updates. In addition to this, the update time allows to derive the number of consecutive omission failures. Whenever T_{IAT} is increased in discrete steps $T_{IAT} = nT_{IAT}, \forall n \in \mathbb{N}$, one or several omission failures have occurred. This holds for both direction uplink and downlink. The generic update time T_{IAT} is defined as follows.

Definition 4.5 (Update time) *The update time T_{IAT} is the time interval from a successful reception of application data message m_i at the consumer until the consecutive successful reception of application data message m_{i+1} at the same consumer.*

$$T_{IAT} = t_{i+1}^{Rx} - t_i^{Rx} \tag{4.5}$$

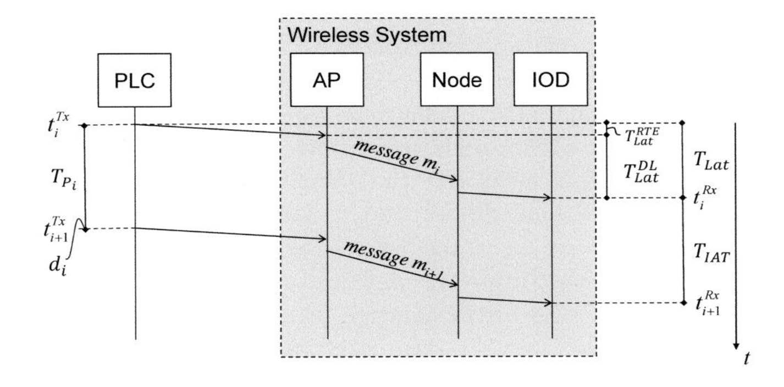

Fig. 4.2: Important metrics: latency T_{Lat} and update time T_{IAT}

The third important metric in the context of simultaneity is the jitter J. The jitter of the update time T_{IAT} is called the rate jitter, the jitter of the latency T_{Lat} is called latency jitter. Both are caused by varying latencies. The jitter is commonly expressed as the standard deviation σ of a given sample. However, this would only be valid if the sample is normally distributed. In this case the metric would cover only 68.2% of the values. In several cases, the difference of the percentile P95 and the percentile P5 is defined as jitter, also commonly used in [143, 172]. For this work the jitter is the difference of the minimum and maximum value and defined as follows.

Definition 4.6 (Jitter) *The rate jitter J_{IAT} is the time difference between the percentile P95 and the percentile P5 of the update time T_{IAT} and the latency jitter J_{Lat} is the time difference between the percentile P95 and the percentile P5 of the latency T_{Lat}.*

$$J_{IAT} = \max(T_{IAT}) - \min(T_{IAT}) \tag{4.6}$$

$$J_{Lat} = \max(T_{Lat}) - \min(T_{Lat}) \tag{4.7}$$

The packet loss rate is also important to be considered, because it is the relevant metric for the reliability of the communication. In this work it is defined from the perspective of the application as follows.

Definition 4.7 (Packet loss rate) *Assuming a set of cyclic real-time traffic flows $F = \{f_1, f_2, ..., f_n\}$ from the application, each transmitted x times. The packet loss rate λ_{PL_i} for a given flow $f_i, 1 \leq i \leq n, n \in \mathbb{N}$ is defined as the ratio of successfully transmitted flows and all transmitted flows.*

$$\lambda_{PL_i} = \frac{x_{f_i^{Tx}} - x_{f_i^{Rx}}}{x_{f_i^{Tx}}} \tag{4.8}$$

Device Model Commonly deployed RTE systems define different roles of devices within their network (cf. Sect. 2.1.1). Two of them are most relevant and further considered in this work, the PLC and the IODs. The PLC is responsible for the execution of the control logic and controls the communication based on the initial configuration of the commissioning phase. It exchanges cyclic real-time traffic flows with one or more IOD, either using the master/slave communication model or the producer/consumer communication model. Throughout this work, only the producer/consumer model is used, but the solution approach is not limited to it. An IOD is a decentralized field device and usually directly connected to sensors and actuators, which provide the interface to the technical process. It is assumed in this work that each wireless node N_i contains one IOD. IODs are configured and parametrised by the PLC and also exchange process data using a cyclic communication. In addition to this, acyclic data such as alarms is exchanged, but these messages are not further considered in this work.

4.1.3 Fault Model

In order to define the constraints under which the system is still able to provide the needed service, i. e., to satisfy the application requirements, a fault model must be established. The fault model specifies what kind of faults can occur and their corresponding effects in terms of failures. In this work the shared wireless medium is considered as the main source of errors, e. g., caused by a decreased link quality, affecting the communication service. Therefore, the fault model for the presented approach in this work considers omission failure semantics, i. e., a message is either lost or received correctly. The following two types of faults, leading to omission failures, are considered and must be handled by the proposed system.

- **Frame loss** - the main reason for this fault is the time- and frequency-variant industrial wireless channel which is mainly influenced by the environment and the mobility of the nodes.
- **Occupied wireless medium** - this fault is caused by uncontrollable transmissions of non real-time nodes in the system. The nodes consist of standard 802.11 technology and are used for monitoring and maintenance.

The first fault is defined to cause inconsistent failures, meaning that it is perceived differently by the nodes. The fault can be further distinguished between the frame loss of a beacon and a loss of a real-time data frame. Since the beacon frame carries several important informations relevant for the wireless network, such as the communication schedule for each cycle, a suitable retransmission strategy is developed to mask the resulting omission failure. For the second case an adaptive retransmission handling is specified. Both measures are of a reactive nature and it is assumed that the omission degree is bounded, because of assumptions for the wireless link, discussed in the next subsection.

The second fault has an omission of the beacon frame as consequence. It is caused by active transmissions of other nodes being not under control of the IWN AP. This is possible, because the system can be deployed in an open environment consisting of other nodes for monitoring and maintenance. The preventing measures in this case are of a proactive nature. They encompass a defined medium usage by the AP before the next beacon frame is transmitted. It requires a tight timing of the transmissions, i. e., the interframe space of two consecutive frames is reduced to the minimal possible time, and a virtual carrier sensing is used. The proposed system can handle a single failure at a time.

4.2 Isochronous Wireless Network

The proposed solution approach of an isochronous wireless network has the main objective to provide real-time communication services for Networked Control Systems satisfying the given application requirements. The approach consists of three main components: the TDMA-based *deterministic medium access control*, the *resource allocation*, and the *provision of a global time base*.

The solution approach is shown in Fig. 4.3. A QoS parameter set represents the application requirements for different traffic flows and is provided to the resource allocation either during startup or when a new node is added to the network. Based on the QoS parameter set, the admission procedure is performed and a new schedule is determined. Both steps consider the current channel conditions and reserve capacity for the recovery of lost frames. The resulting schedule is passed to the TDMA-based medium access control which assigns time slots to the nodes using the global time base. Moreover, the global time base allows the integration into existing wired networks due to a synchronous communication interface.

The specified components of the approach are addressing the identified problems in Sect. 1.2 in order to enable the IWN to satisfy the application requirements (cf. Table 1.2). This section introduces the main conceptual components, their interactions as well as the specific requirements addressed by them.

The deterministic medium access control and its relation to the other components is introduced in Sect. 4.2.1. A detailed component description with an analysis of its main characteristics is provided in Ch. 5. An overview of the resource allocation is given in Sect. 4.2.2, and its detailed description follows in Ch. 6. The provision of a global time base and its relation to the other components is discussed in Sect. 4.2.3. This component is described in Ch. 7 with a specific focus on the more challenging part of wireless clock synchronization and an analysis of relevant characteristics.

Finally, it must be considered that a wireless system in industrial automation is always connected to a wired real-time network [150, 127]. Even though the pre-

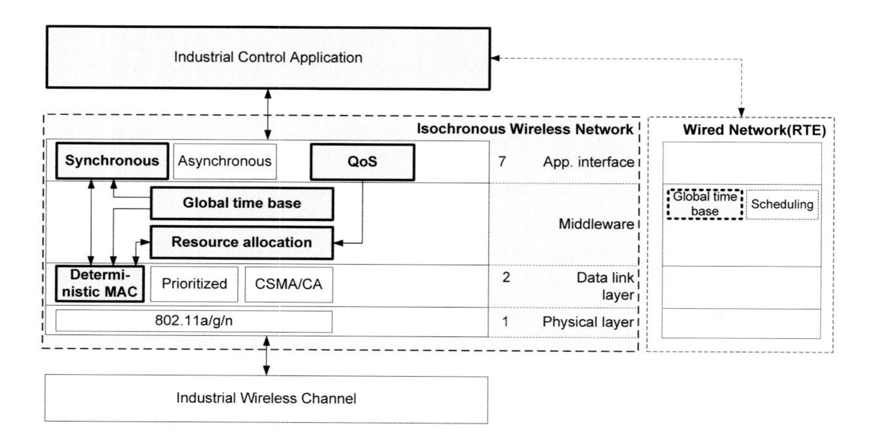

Fig. 4.3: Isochronous wireless network for industrial automation

sented approach and the developed conceptual components focus on the wireless part, the wired real-time network and its properties must be considered sufficiently in order to identify challenges that are related to the integration of the wireless solution into existing wired real-time networks. The resulting integration aspects for a hybrid wired and wireless system are discussed in Sect. 4.3 in terms of mapping strategies and the resulting implications on the application.

4.2.1 Deterministic Medium Access Control

The first component of the approach is responsible for providing a deterministic medium access. This component provides mechanisms, to avoid unpredictable latencies, introduced by uncoordinated distributed MAC protocols. It is designed with the objective to be efficient for cyclic process data frames with small payloads. The deterministic medium access control component is also responsible for detecting, and preventing faults according to the fault model, described in Sect. 4.1.3.

This component consists of two main building blocks, the Isochronous Medium Access Control (IsoMAC) and the frame error recovery.

IsoMAC is based on a TDMA scheme for the coordinated medium access of all nodes (cf. Sect. 5.2). Therefore, the IsoMAC organizes the communication cyclically. Each cycle is subdivided into a scheduled phase and a contention phase. Every cycle starts with a beacon frame, which distributes the communication schedule information and the clock synchronization information for establishing a global time base to the nodes.

The real-time traffic flows are transmitted in the scheduled phase using assigned timeslots according to the schedule and the global time base. The transmission of real-time traffic flows is separated by their direction, i. e., first the downlink frames and then the uplink frames (see Sect. 5.2.2), because this allows temporal optimizations when integrated with a wired network.

Best effort frames and management frames, e. g., for clock synchronization, are transmitted during the contention phase. The scheduler always ensures that at least one best effort frame with a maximum payload size of 2304 Byte can be transmitted within each cycle. To prevent an uncontrolled access of standard wireless nodes, e. g., used for maintenance, and the resulting unpredictable behaviour, the real-time transmissions are separated by the shortest possible interframe space (SIFS) and protected by a virtual carrier sensing. A restricted phase at the end of the cycle protects the next beacon frame transmission (Sect. 5.2.3).

Furthermore, the global time base is used by this component for synchronization of the wireless cycle to the communication cycle of the wired network. This avoids asynchronously operating applications and their major drawback of an increased latency. This is further discussed in Sect. 4.3.

Frame Error Recovery The frame error recovery detects whether a transmitted frame must be retransmitted or not. For the downlink direction, the error detection is done through acknowledgements embedded within the uplink traffic. For the uplink direction, the error detection is handled by the AP (cf. Sect. 5.3), in order to be more efficient for small payload frames and to reduce the protocol overhead. Even though retransmissions ensure an increased reliability, the communication jitter is negatively influenced and increased. A trade-off between retransmitting a frame and the resulting jitter must be made depending on the technical process. Thus, the frame error recovery is closely linked to the adaptive retransmission handling of the resource allocation component described in the next subsection.

4.2.2 Resource Allocation

The resource allocation component is responsible for allocating the required resources for real-time traffic flows. It considers currently admitted traffic flows of the AP and the available capacity of the channel. It also guarantees that the application requirements can be met by all admitted flows. This component addresses the timeliness and simultaneity requirements, because the problem of an unpredictable behaviour of the wireless network due to an uncontrolled increased utilization and the resulting overload situation in the network, referred to as congestion, is avoided. The component consists of three building blocks, the admission control, the scheduler, and the adaptive retransmission handling.

Admission Control The admission control is mainly dealing with the admission procedure of new real-time traffic flows based on their requirements as specified in the QoS parameter set (cf. Sect. 6.2). The admission control is based on the specified requirements of a new flow, the current utilization of the system and the status of the wireless resource. It also considers priorities defined by the application to be able to reject flows with a lower priority, if needed. It is implemented using a pessimistic approach, because it allows to transmit each flow twice. The available resources must always allow the dynamic retransmission handling to guarantee the needed reliability for the application. Based on these pessimistic assumptions, the admission control performs a schedulability test. If the flow can be scheduled, the new flow is admitted, if not, the requesting node is informed about the rejected request. Even though the admission control is pessimistic, the allocated bandwidth for recovery can be used by best-effort traffic when no retransmissions are needed.

A cross-layer frame inspection might be used at the application interface to automatically collect the application requirements, which are needed for the flow admission procedure. The requirements are also defined as tuples consisting of the relevant attributes, such as update time, and omission degree. The cross-layer inspection is also used for a continuous monitoring of the application traffic flows and provides the obtained information about the omission degree of the monitored flows as input to the dynamic retransmission handling, which is able to react depending on the retransmission policy.

Scheduler The real-time communication properties of the network are based on the TDMA scheme that is combined with a specifically designed scheduler within the AP. Together, both components provide mechanisms to dynamically allocate bandwidth, reschedule the communication in the scheduled phase, if necessary, and to distribute the new schedule to all nodes with real-time traffic flows. The scheduling approach is based on the Earliest Deadline First (EDF) scheduling algorithm (cf. Sect. 6.3). The deadline of a synchronous real-time traffic flow f_n is assumed to be set to the next relevant cycle T_{IAT} of the flow. The aim of the scheduler is to produce a feasible schedule, i. e., the deadlines of all admitted real-time traffic flows have to be met. The schedule is sent to the AP and cyclically transmitted within the beacon frame.

Adaptive Retransmissions The adaptive scheduling of retransmissions allows an optimization of the jitter in the presence of frame errors (see Sect. 6.4). Based on the frame error information of the medium access control component an adaptive retransmission scheduling is done. Depending on the application requirements, it enforces different retransmission policies. Either with the objective to optimize the PLR and achieve an increased reliability for the transmission or to optimize the isochronous characteristics of the wireless system.

The selection of the policy is always a trade-off between the isochronous characteristics of the protocol and having a high reliability of frame transmissions even when the wireless channel quality degrades. The more important characteristic must always be determined depending on the relevant application and the environment.

4.2.3 Global Time Base

The global time base is responsible for two aspects. First, the wireless synchronization allows to synchronize the clocks of the wireless nodes with the required accuracy. This is a basic requirement for the TDMA-based medium access control.

Secondly, the synchronization approach allows a seamless integration into the wired network to avoid an asynchronous and decoupled operation of both systems.

The wireless synchronization concept is based on embedding the synchronization information into the IsoMAC beacon frames and add additional redundancy for increasing the reliability (cf. Sect. 7.2). This approach considers the limited capacity of the wireless channel and results in an increased efficiency as compared to sending separate frames for clock synchronization.

To ensure real-time communication with minimal latencies for the whole system, there is a need to provide temporal consistency to enable a synchronous communication between the wired and the wireless system. The seamless integration of both systems is ensured by using a solution based on IEEE 1588v2 [68], widely accepted in RTE networks, for providing a global time base. The wired system supports hardware timestamping resulting in a synchronization accuracy of a few nanoseconds, whereas the wireless system mainly affects the achievable synchronization accuracy. The clock synchronization over the wireless network can not achieve the same accuracy as its wired counterpart, mainly because of the software timestamping as error source and a lacking hardware support. Thus, the developed modifications of this component focus on the wireless part of the system.

4.3 Integration Aspects

In order to design a hybrid system consisting of the IWN, which is integrated into an existing wired RTE network, it is necessary to consider two aspects, existing integration strategies and possible implications of them.

Cena et al. propose in [20, 21] several approaches for RTE extensions with wireless networks. They have clearly identified that the integration of any wireless network would have consequences for the tight timing of the RTE. Hence, in this work a proxy concept is introduced. The concept allows a decoupling of the wired RTE with its very tight timing and the IWN without compromising the advantages of having both systems synchronized with a global time base. The proxy concept is shown in Fig. 4.4 and consists of the IWN AP, a synchronized mapping component, and 1...n virtual IODs actually representing the wireless nodes. The latter depends on the chosen mapping concept.

4.3.1 Mapping Concepts

As an example, the mapping concept of Profinet is briefly introduced. The described integration is not limited to Profinet and also applicable to other RTEs, because they provide similar approaches. A possible mapping strategy always depends on the device model of the RTE and the number of possible mapping levels, such as slots and subslots for the Profinet example.

In Profinet three different mapping concepts are possible. They are mainly arising from work in the area of field bus integration into RTE solutions [141], but the concepts allow an extension for the integration of wireless networks [165]. The

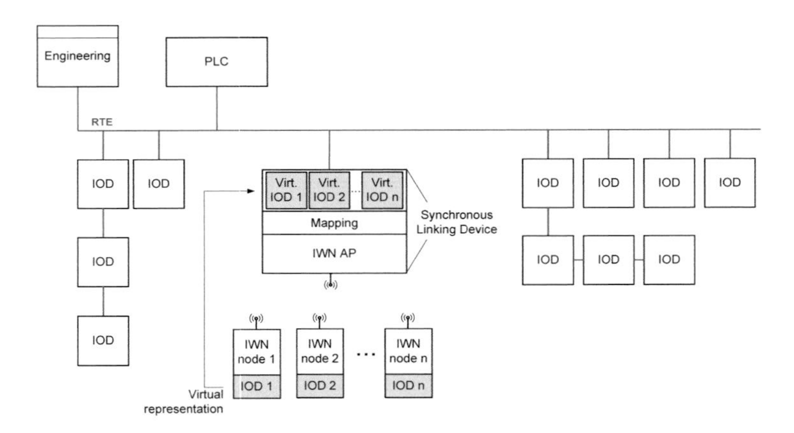

Fig. 4.4: Integration of the wireless system into existing real-time networks

compact mapping approach maps the whole wireless network with all nodes to a single IOD, having one slot and at least one subslot. Using *transparent mapping* every wireless node is mapped to a single IOD within one synchronous linking device following the multi-device approach described in [76]. This mapping concept is shown in Fig. 4.4. The *modular mapping* concept maps every wireless node in its own corresponding module or submodule in an IOD, according to the slot/subslot approach [76].

All mapping concepts have in common, that all wireless nodes must be included during the commissioning phase a priori due to the static engineering and scheduling approaches of RTE systems. This might cause problems if certain wireless nodes are not always part of the wireless network. For instance, a reconfigurable manufacturing system frequently changes, because modules of the manufacturing system are added or removed due to its nature. Promising solution approaches are discussed by Dürkop et al. in [32] and Wisniewski et al. in [183]. However, these new plug&work mechanisms for industrial automation systems and the arising engineering challenges of such a flexible system are not within the scope of this work and the normal static nature of RTEs is assumed.

4.3.2 Hybrid System Implications

The integration of the wired and wireless system addresses the problem of asynchronously operated networks. If both networks would be not synchronized, i. e., the communication would be asynchronously, the temporal behaviour in terms of latency is significantly degraded. The problem of asynchronously running communication systems is well-known in the literature. For instance, it is shown by Höme et al. in [53] that the concatenation of different communication systems with different asynchronous cycles have a very bad impact on the overall temporal behaviour of the system. Usually, this can be neglected, if $T_{Cycle} \gg T_{App}$, where T_{Cycle} is the cycle time of the wired communication system and T_{App} is the control cycle of the application. For the given application scenario this is not possible for the wireless network, because of its maximum possible cycle time $T_{NUT} \approx T_{App}$. Thus the available global time base must ensure a system wide synchronization of both communication cycles and all involved nodes.

In addition to this, specific application synchronization requirements are important and further discussed here. As defined in Sect. 4.1.2, the implementation of an NCS requires that all sensor values are read in at a defined time t_{in} with the application cycle time T_{App} to provide a system wide consistent image of the process to the PLC. The same holds true for the output of the PLC to the actuators at t_{out}.

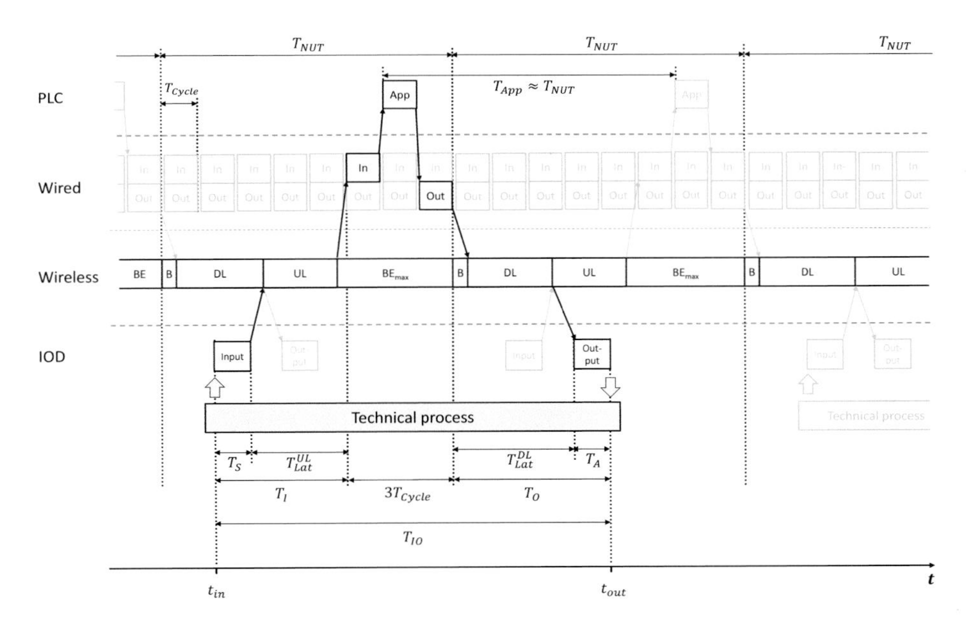

Fig. 4.5: Isochronous communication network for a hybrid wired and wireless infrastructure

Besides the application cycle time T_{App} the achievable response time T_{IO} is important for the NCS. The dependencies are shown in Fig. 4.5. The state of the technical process is read at t_{in} by all sensors. Then the values are transmitted from the IOD to the PLC via both networks. The PLC processes the data and calculates new actuator values. They are sent back from the PLC to the IOD and provided at t_{out} to the actuators. The resulting response time can be described with Eq. (4.9).

$$T_{IO} = T_I + 3T_{Cycle} + T_O, \tag{4.9}$$

with

$$T_I = T_{Lat}^{UL} + T_S, \tag{4.10}$$

$$T_O = T_{Lat}^{DL} + T_A, \tag{4.11}$$

where T_S and T_A are processing times of the sensor and actuators. Since they are comparably small, i. e., $T_S \ll T_{Lat}^{UL}$ and $T_A \ll T_{Lat}^{DL}$, it can be assumed that T_I and T_O are mainly determined by the wireless latencies.

5

Deterministic Medium Access Control

In this chapter an approach for a deterministic medium access control for industrial real-time traffic is proposed. The determinism is achieved by a central coordination of the medium access using a TDMA-based approach. According to [19] this is only solution to avoid random delays when accessing the shared wireless medium. The presented approach is using appropriate mechanisms on layer 2 of the communication system to achieve this goal.

Different challenges arise for the design of the TDMA-based approach and must be addressed accordingly. First, the efficient transmission of real-time traffic with small payloads is very important, because of the given traffic characteristics as defined in Sect. 3.1. At the same time a sufficient capacity for high throughput applications must be provided without providing any QoS guarantees. Second, the dynamic properties of the wireless channel must be considered in order to achieve a sufficient reliablity. And third, the integration of existing standard nodes without any influences on the remaining real-time traffic must be possible.

5.1 Coordinated Medium Access Control

Within this section the related work for medium access control approach in wireless networks is discussed and analyzed with respect to the given requirements and the aforementioned challenges imposed by the application. Since Unified Modeling Language (UML), as state of the art protocol description language, will be used within the protocol definition, it is briefly introduced in this section.

5.1.1 Related Work

The related work is further divided in two different categories, (i) generic approaches to modify the existing 802.11 medium access to achieve optimizations for various application domains, mostly voice and video transmission, and (ii) research works regarding coordinated medium access control specifically conducted in the context of industrial applications.

Approaches for various application domains The authors of [158] specify a simplified polling scheme which avoid transmissions when a station has no frames to transmit. The polling period can be reduced if a station request is sent. The maximum polling period is defined by the user depending on the requirements. An adaptive polling scheme is proposed by Milhim and Chen in [117] using two dynamic polling lists to reduce the medium access delay and the polling overhead. The polling overhead can be significantly reduced, and a higher throughput can be achieved as compared to the ordinary Round-Robin scheme.

Lo et al. [104] extend the DCF medium access with polling using an efficient multipolling mechanism. It is called CP-Multipoll and different backoff values are used for different traffic flows to establish a polling list. Each station must content for the medium when it receives the multipoll frame. Due to the polling list the order for contention is assigned before. The transmissions are started immediately after a multipoll is received to avoid a contention with legacy stations operating with DCF.

A Soft-TDMAC for mesh networks is proposed by Djukic et al. in [29]. Even though a TDMA-based protocol is proposed and the approach seems to be interesting, it does not provide many details and the work mainly focusses on clock synchronization aspects in mesh networks. It is shown that the proposal achieves accuracies around 20 µs which is not sufficient for this work (cf. Ch. 7).

All described proposals have in common that they are designed for variable and constant bit rate traffic without considering properties such as isochronous transmission. The variation of update times and latency is not considered as long as the traffic adhere a given deadline. Since the proposals are optimized for guaranteeing a defined throughput, frames with very small payloads are less important and the arising challenges are not explicitly considered.

Approaches for industrial applications It has been highlighted in [182, 35] that industrial WLAN deployments are currently limited to applications with relaxed temporal requirements such as monitoring and diagnosis, because IEEE 802.11-based WLANs in their current state are still not deterministic and less reliable than wired networks. Therefore, concepts related to and industrial WLANs have been already addressed in several other research works. At the same time both works identified the IEEE 802.11 as a very promising technology for several application areas, if they are extended to provide QoS guarantees as required by industrial applications.

Nett et al. propose in [130, 154] a four layer middleware for real-time communication in the context of cooperating robots. The necessary real-time requirements of the application could be met by means of the middleware between the data link layer and the application layer. Using this approach, a reliable communication could be established and end-to-end delays of approx. 40 ms are achieved which belong to real-time class 3.

Moraes et al. propose in [124, 123] a very interesting extension of the the normal EDCA. It was specifically tailored for industrial usage, fulfilling the temporal requirements of real-time class 3. In [179] Willig proposes a flexible TDMA (FT-

DMA) which is based on a polling mechanism, i. e., a coordinating station explicitly informs all involved real-time nodes about the schedule and causes additional overhead. Specifics about the physical properties of industrial channels and corresponding conclusions with respect to the design of MAC and link-layer protocols are proposed in [180].

Several works deal with issues related to hybrid wired/wireless networks [1, 21]. For instance, Seno et al. [156] focus on a wireless extension of Ethernet Powerlink based on 802.11, but without any modification of the wireless system. All of the mentioned works are using standard technology and optimize it according to their needs by modifying accessible parameter of the standard system, i. e., these works mainly focus on a prioritization scheme rather than a coordinated medium access.

A TDMA-based system using a higher layer coordination of the medium access is proposed in [26, 175]. Even though the work is done in the context of industrial applications, the proposal is not considering a full integration into wired communication systems, i. e., neither a synchronization to the wired infrastructure nor to the application are reflected in the work.

A framework for reliable real-time communication was applied to the IEEE 802.11 standard in [96]. It is based on a new retransmission scheme residing on the transport layer, but also uses polling as medium access. The main focus of this work was put on reliability of real-time communication, but only fulfilling requirements of real-time class 3.

In [5] a wireless extension of FTT-Ethernet [138] called wireless flexible time triggered protocol is investigated. The protocol is based on a medium access determinism using a bandjacking technique by means of a wideband interferer. This is very critical in terms of coexistence, especially in industrial environments where this is of major concern.

The authors of [6] propose a wireless gateway concept for a wireless point to point connection and a Sercos III system. Even though the authors promise to focus on wireless factory control applications with isochronous real-time requirements, the presented concept just uses an IEEE 802.11n card without any modification, i. e., the standard CSMA/CA is used as medium access control.

In order to further increase the temporal characteristics and reliability of wireless transmissions the works [145, 22] propose to use a redundant channel. The approach can be helpful in harsh environments, but it can not solve the problem of having undeterministic medium access delays in distributed wireless networks.

Discussion Summarizing the main results of the previous analysis of related work in the area of coordinated medium access control, it can be concluded that several interesting proposals exist, but none of them fulfils the requirements of this work. The efficiency of the protocol extensions in the context of industrial traffic is never considered. Even though several works are aiming at a hybrid wired/wireless network architecture, these works only apply existing technologies. A full integration including the required modifications of the wireless network, especially in terms of suitable mechanisms for medium access and global time base, has never been applied.

5.1.2 Protocol Description using UML

In order to describe communication protocols in a formal way, different formal description techniques (FDT) exist, such as the Specification and Description Language (SDL) or Message Sequence Charts (MSC). Even though UML does not belong to FDTs, it is extensively used in the recent years to describe protocols and its usage will probably increase in the future [91]. UML is a modeling language for software and system development. The trend of using UML is because of its simplicity, and being close enough to the major concepts encountered in most programming languages. For instance, when a protocol model is created and implemented, it is very difficult for a third party to understand the concepts if only the source code is available. Using UML as a semi-formal description or to create a model independent from a programming language ease the understanding of it, especially if working in interdisciplinary teams. One of the biggest advantages of UML is its flexibility and that it can be used not only to create models for documentation or specification, but also to simulate the model and directly generate deployable code out the information. Therefore, UML is chosen to describe the proposed IsoMAC protocol in this chapter.

As extensively described in [91], two UML diagram types are relevant for specifying communication protocols. The *state machine diagram* is used to model finite state machine representations of different protocol entities. Thus it is a behaviour-oriented view of the protocol entities. Whereas UML *sequence charts* are used to model the protocol message exchange between different protocol entities. This can be considered as the communication-oriented representation of a protocol [87].

5.2 Isochronous Medium Access Control

The concept of a new IsoMAC protocol is proposed in this section. The overall MAC architecture of IEEE 802.11 including the proposed IsoMAC is shown in Fig. 5.1. The protocol considers the specific characteristics of industrial traffic, as discussed in 3.1, most importantly the cyclic behaviour and the small payload sizes. For real-time critical industrial traffic either a time division multiple access (TDMA) based mechanism has to be used or it has to be based on polling (e. g., HCCA) [19].

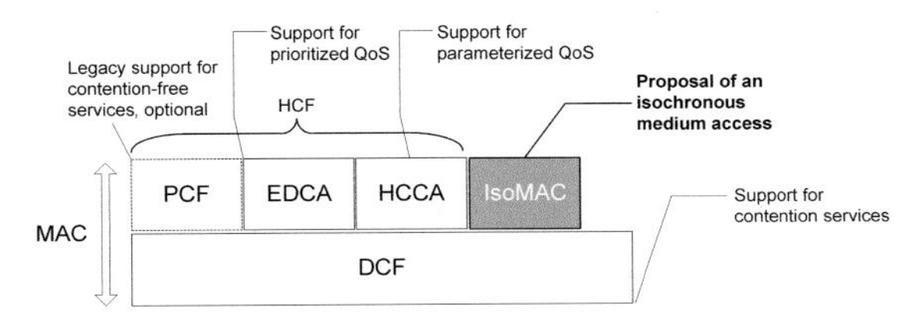

Fig. 5.1: IEEE 802.11 MAC architecture with the proposed IsoMAC

As discussed in the previous section, existing approaches using HCCA or modified versions of it as well as other polling-based approaches have some drawbacks and weaknesses. Especially, when transmitting frames with a small payload, the necessary polling messages result in a huge communication overhead [158]. Since the additional protocol overhead results in a deterministic, but inefficient medium access and a reduced scalability of the system, it is more promising to use a TDMA like scheme [19]. Hence, the IsoMAC proposal is based on a time division multiple access (TDMA). It can coexist with standard 802.11 stations for monitoring, i. e., it supplements the existing medium access mechanisms by means of a deterministic and efficient protocol. Moreover, it is specifically tailored for usage in industrial applications and their needs.

5.2.1 Network Topology

The network topology for an IsoMAC enabled wireless communication system is restricted to infrastructure networks [70]. An example topology is shown in Fig. 5.2. The wireless network can be considered as a wireless extension of an existing Real-time Ethernet (RTE). The RTE consists of common components, such as an IO Controller (IOC) as controller, and several IODs as interfaces to sensors and actuators. Whenever a wired connection is impossible due to mobility demands, additional devices can be wirelessly connected via the IWN including IsoMAC. It consists of an IWN AP and several wireless nodes, either with or without IsoMAC.

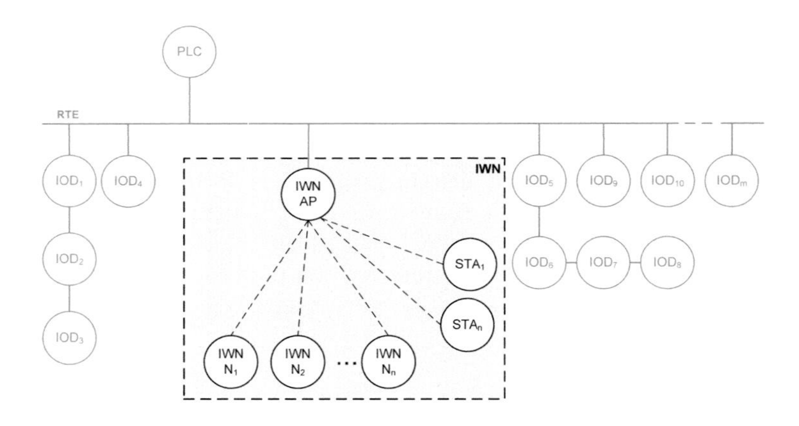

Fig. 5.2: Network topology for the IsoMAC

The IWN AP is considered as the centralized controller and manager of the wireless network. It provides the interface to the wired RTE network and is in charge of coordinating the wireless network. The most important task is to coordinate the medium access, in order to achieve a deterministic behaviour and to assure negotiated real-time requirements. Furthermore, the AP is responsible for accepting or rejecting wireless nodes which are requesting certain resources. It is also in charge of dynamic resource allocation, as discussed in Sect. 6 depending on the current number of accepted traffic flows.

The wireless IWN nodes $N_1 ... N_n$ provide the interfaces to the application. They are able to communicate in a deterministic way and to use the medium access which is specified later on in this section. If a new IWN node is added to the wireless network, it has to establish the needed traffic flows immediately by means of a resource reservation procedure with the IWN AP.

Finally, standard nodes $STA_1 ... STA_n$ are also permitted to get associated to the IWN AP. They can not request any resources beforehand, and the accomplishment of real-time requirements are not guaranteed. Standard clients have to use a standard coordination function, such as DCF or EDCA as defined in [70]. These nodes are mainly used for monitoring purposes, and for other common tasks which do not require real-time communication.

Currently, the proposed IWN is based on 802.11g (2,4 GHz-ISM-band), 802.11a (2,4 GHz-ISM-band), or 802.11n as its physical layer (PHY), but in general applicable to any wireless network. The coverage of the network depends on the chosen frequency and the characteristics of the industrial environment, but certain assumptions are taken regarding the availability of the wireless resource as part of the network model in Sect. 4.1.1.

5.2.2 Coordinated Channel Access

The IsoMAC protocol basically includes resource reservation along with admission control, scheduling and medium access. It provides mechanisms to dynamically allocate resources, reschedule the communication in the planned phase and to distribute the new schedule to all involved nodes.

A common approach for a suitable resource reservation and admission control procedure is specified in [70]. The IsoMAC procedure will be mainly based on it. Before an IsoMAC node associates to the AP, the necessary resources for a traffic flow f_i have to be specified by means of a resource request frame which includes a QoS parameter set described as tuple $Q_i = \langle T_{P_i}, d_i, OD_i, l_i \rangle$, with send period T_{P_i}, deadline d_i, the maximum omission degree OD_i^{max}, and payload size l_i, as introduced in Sect. 4.1. The admission control of the AP uses the traffic specification to process whether the new node can be accepted or not. The decision depends on the already admitted traffic flows and the corresponding available resources for the IWN. The admission control and scheduling approach is considered in the next chapter. The medium access mechanism will be discussed in the following subsections.

The main design objective for the IsoMAC medium access was the ability to provide a deterministic behaviour and to support industrial soft real-time traffic flows. The resulting channel access mechanism is shown in Fig. 5.3 and is based on a TDMA scheme. One communication cycle consists of different phases, a scheduled phase (SP) for process data, i. e., real-time traffic, and a contention phase (CP) for

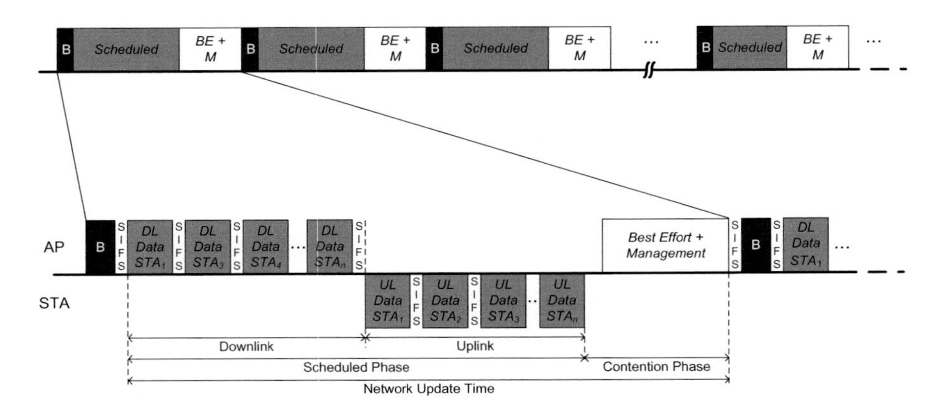

Fig. 5.3: IsoMAC channel access

best-effort (BE) and management (M) traffic. The behaviour of both existing protocol entities is shown in Fig. 5.4(a) for the AP (or coordinator) and in Fig. 5.4(b) for the node, whereas the communication oriented representation of the protocol is shown in Fig. 5.5.

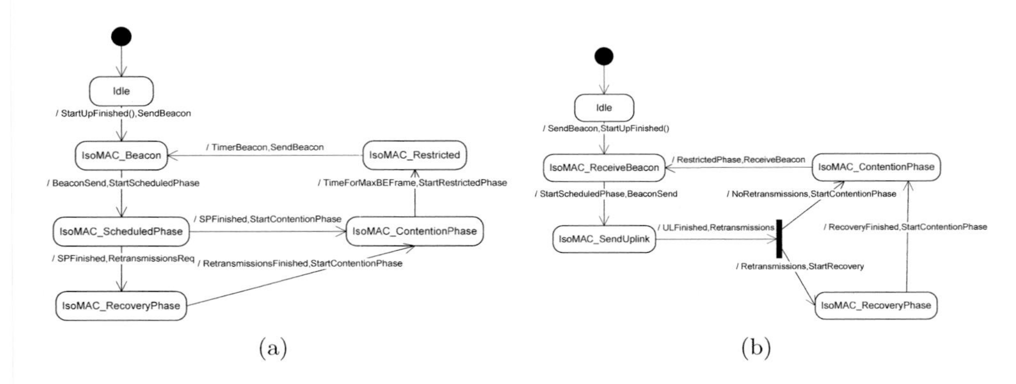

Fig. 5.4: UML state machine diagram for the protocol entities AP (a) and node (b)

Real-time process data is transmitted by the IsoMAC nodes within assigned timeslots during the scheduled phase. The SP is further divided in timeslots for downlink (DL) traffic and for uplink (UL) traffic. The first part is reserved for DL real-time data, i. e., from the AP to the nodes. The second part is for UL real-time data from the nodes to the AP. The repetition interval of the scheduled phase is called NUT T_{NUT}, since it is equal to the shortest possible update time of the nodes.

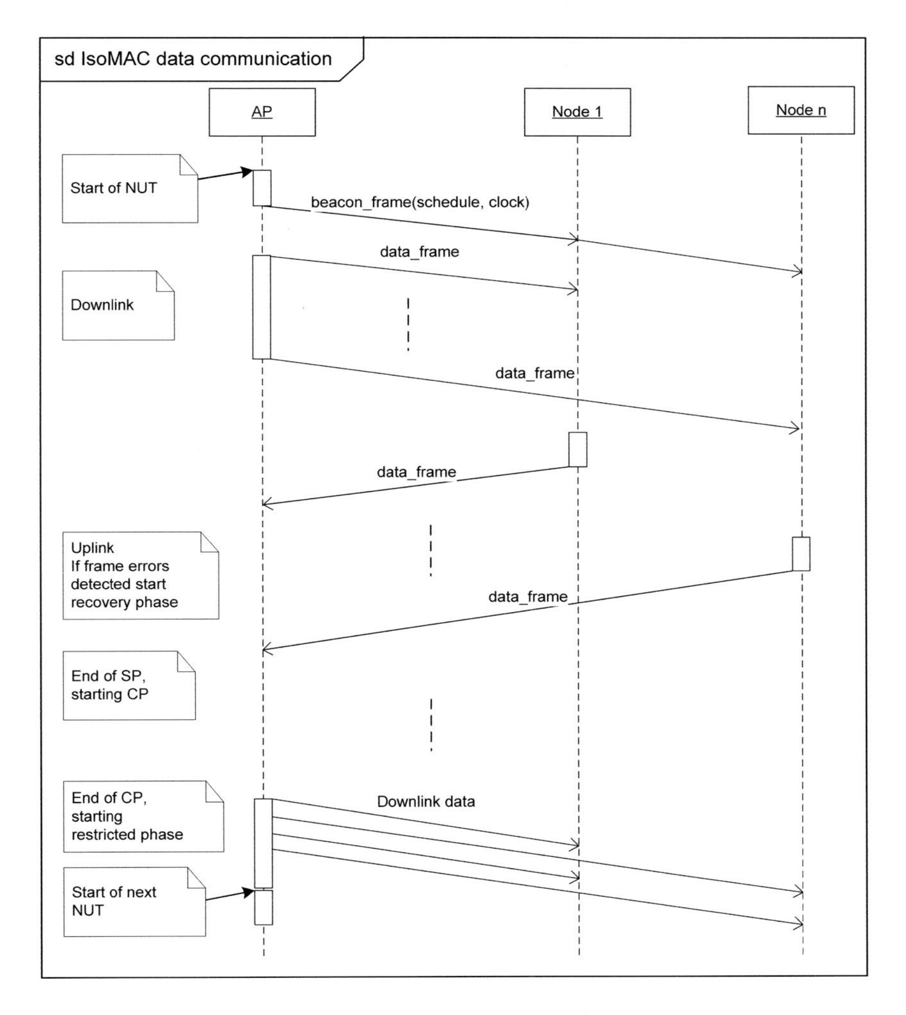

Fig. 5.5: UML sequence diagram of the IsoMAC data communication

Best effort (BE) traffic and management frames, such as messages for clock synchronization, are transmitted during the contention phase. During the contention phase the ordinary DCF and EDCA are used to contend for the medium, i. e., standard 802.11 stations can coexist and exchange data during the CP. The traffic scheduler always ensures that at least one data frame with a maximal sized payload of 2304 Byte can be transmitted within the CP. The BE frames are handled by using a standard first in first out (FIFO) queuing strategy. The strategy also implements a mechanism to assign a maximum live time to all frames within the queue t_{TTL} depending on their time of arrival. It discards every BE frame which exceeds the defined threshold t_{TTL} in order to avoid a congestion situation for the BE traffic.

5.2.3 Schedule Distribution and Restricted Phase

Every beacon frame contains the schedule information for the nodes, i. e., once per beacon interval the dynamic schedule can be changed and distributed accordingly. Thus, one schedule information element (SIE) per traffic flow is included in the beacon payload. The start time of the transmission, a unique traffic flow identifier (TFlowID), and the station identifier (StationID) is contained in every SIE.

Due to the schedule distribution mechanism, the beacon frame is of vital importance for the isochronous MAC protocol. Hence, delaying a beacon frame due to a prolonged CP would cause severe problems. This issue has been identified in [95] with the conclusion that such a situation highly degrades the real-time behaviour. The transmissions within the CP are rather unpredictable, which might lead to an undesired foreshortened Scheduled Phase (SP) during the next NUT. Since all BE stations are allowed to transmit frames at any time during the CP, it might happen that the medium is still busy at the time when the next beacon is supposed to be transmitted. This would severely affect the provided QoS, as unpredictable delays at the beginning of a SP are introduced and the whole SP will be shifted.

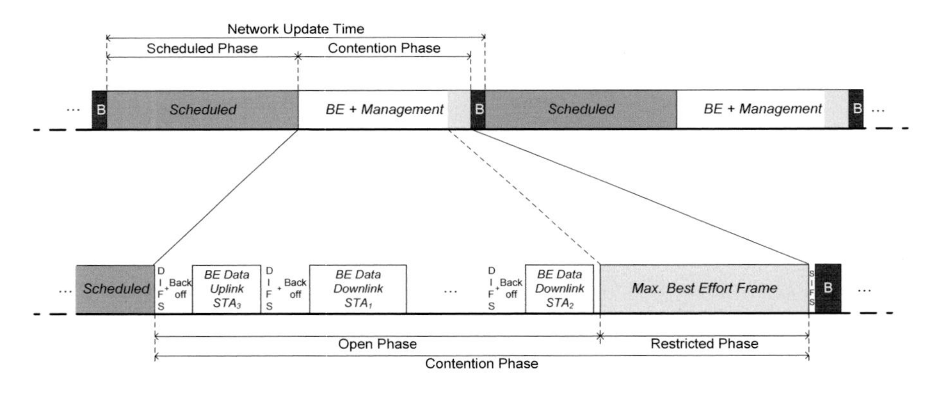

Fig. 5.6: Prevention of a foreshortened scheduled phase

Thus the beacon should be always transmitted at equidistant time intervals equal to the required NUT. In order to achieve this, the CP is further divided into an open phase and a restricted phase as shown in Fig. 5.6 and Fig. 5.4(a). During the open phase all nodes and legacy stations are allowed to content for the medium, whereas the restricted phase has always the length of the duration of a maximum frame transmission. Whenever no transmission is initiated at the beginning of this phase, the AP undertakes control of the medium again, preventing undesired Best Effort (BE) transmissions. In case a transmission starts, once it has been completed, the AP starts to control the medium afterwards, no matter whether it was a maximum frame or not. This allows a predictable termination of the CP without having any knowledge about traffic characteristics of the BE stations.

5.3 Frame Error Recovery

Since the IsoMAC proposal is based on the IEEE 802.11 standard [70], it uses a modified approach of MAC layer based acknowledgements for unicast communication to increase the communication reliability. The adaptive retransmission handling is closely linked to the channel access, the scheduling strategy and the application. The corresponding retransmission strategies are very important and consider the entire collection of real-time flows being served by the network, while also maximizing usage of network resources.

Therefore, the IsoMAC frame error recovery procedure is designed efficiently, in order to reduce the protocol overhead. The transmission of acknowledgment frames (ACKs) for the reception of downlink frames is postponed until the uplink sub-phase. Conversely, uplink data frames are not explicitly acknowledged by the AP. The reason for the latter is that the AP takes ownership of error recovery and uses its knowledge of the schedule to determine when frames should have been received to detect if a communication error has occurred. In order to deal with communication errors the scheduled phase is further subdivided and a recovery phase is added. The recovery phase will immediately take place after the scheduled phase is finished. During the recovery, frame retransmissions of downlink and/or uplink frames can take place if necessary.

5.3.1 Recovery Phase

In order to deal with omission faults the scheduled phase is further subdivided and a recovery phase is introduced. The recovery phase will take place after the first downlink and uplink scheduled periods are finished. The duration of the recovery phase depends on the number of transmission errors occurred during the scheduled phase and will be considered by the admission control procedure. Additionally, aperiodic real-time traffic can be served during the RP.

In order to allocate the added resource needed for the RP, the duration of the CP will shrink accordingly. Depending on the application, a minimum duration for CP is respected in order to secure resources for best effort and management traffic within every NUT. Hence, the maximum length of the recovery phase T_{RP}^{max} is bounded. In order to guarantee determinism and repeatability of the cyclic real-time traffic, the length of the NUT T_{NUT} is fixed and its duration is determined by the minimum service interval required by the application. The duration of the scheduled phase T_{SP} depends on the number of admitted flows and the amount of data per flow. Therefore the maximum duration for the recovery phase is given by:

$$T_{RP}^{max} = T_{NUT} - T_{SP} - T_{CP}, \tag{5.1}$$

where $T_{CP}^{min}$is the minimum allowed CP duration. Note that the recovery phase does not take place for those NUTs, where there were no communication errors, thus maximizing the CP length. Also, because the RP occurs after the SP, retransmissions do not impact the delay performance of any real-time traffic flows, neither in the present nor in subsequent NUTs.

5.3.2 Beacon Recovery

The beacon frame is of vital importance for the whole protocol. Since broadcast and multicast frames are unacknowledged, it is not possible to know if the beacon frame which includes the schedule has been successfully received by every node that needs to communicate during the scheduled phase of the NUT. When a node fails to receive the beacon, it will remain silent when its transmission opportunity arrives. This situation will leave the channel idle for, possibly, long enough (DIFS) so that other BE STAs attempt transmission. Hence, the remaining of the scheduled phase may be spoiled, due to unnecessary contention for the medium or collisions. It is therefore necessary to implement a schedule distribution recovery mechanism, to allow the scheduled node to transmit its uplink traffic and also to avoid communication errors due to interference from BE STAs.

In IsoMAC, downlink communication precedes uplink traffic. Therefore, it is not possible to know if a node successfully received its SIE until its TXOP (transmission opportunity) arrives and it can send out data and either an ACK or a BACK. Once a AP realizes that such situation has occurred, it will trigger a recovery mechanism.

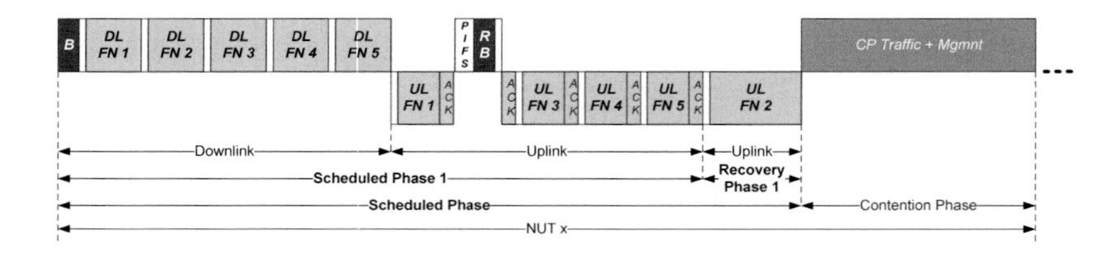

Fig. 5.7: Beacon recovery

The proposed mechanism is shown in Fig. 5.7. If the AP senses the channel as idle for a PIFS, then it takes control of the medium and transmits a recovery beacon frame (RBeacon) containing the remaining SIEs only. This increases the chances that nodes that missed the first beacon, receive it this time. Waiting a PIFS ensures that no BE STA win access to the medium until the channel has been idle for longer than DIFS, i. e., PIFS $<$ DIFS). If there is enough time left, the node that was supposed to transmit its uplink traffic can do so during the remaining time. The uplink traffic could be either data, or ACK/BACK, or both if time permits. The transmission duration of the RBeacon frame must fulfill the constraint $T_{RBeacon} \leq T_{TXOP_i} - \text{PIFS}$, where T_{TXOP_i} corresponds to the original uplink TXOP of the *ith* node. Fig. 5.7 shows the case, where there was enough time to transmit a RBeacon and a ACK frame. The RBeacon provides an additional TXOP for node 2 at the end of the PSP, thus leaving the rest of the original uplink schedule unchanged. In this example, the RP begins with the transmission of uplink data from node 2. Afterwards, if no errors occurred during transmission, the CP begins. In case resources can not be allocated to create a RP, node 2 can attempt to transmit its data within the CP using the highest EDCA priority.

Regardless of what time is left before the next TXOP, the AP should remain in control of the medium to prevent the channel from being accessed by BE nodes. If this is not possible, the SP is still protected by the virtual carrier sensing mechanism using the network allocation vector (NAV). For instance, if there is plenty of time, and the RBeacon was not received by the node, the original process (waiting PIFS after sending an RBeacon) can repeat itself until successful delivery. The time left and the response (or lack thereof) from the node determines the course of action (re-sending RBeacon, transmitting uplink data, transmitting only ACK, etc.). As long as inequality $T_{TXOP_i} \leq \text{PIFS} + T_{RBeacon} + \text{SIFS} + T_{Ack}$ holds, it can be guaranteed that the full beacon can be retransmitted and that the node manages to acknowledge both, the downlink traffic it received and the new beacon frame. In this way the AP would be able to reschedule not only the uplink but also downlink of the incumbent node in case it is necessary.

A procedure similar to the one described above, is used to recover downlink and uplink omission failures. The mechanism belongs to the adaptive retransmission handling and is described in the next chapter about resource allocation.

5.4 Analysis of the Isochronous MAC

The proposed isochronous medium access control is analyzed in this section. Several relevant characteristics are determined and compared to the existing controlled HCCA medium access of the 802.11 standard, which can be considered as state of the art. The analysis highlights the obtained improvements of the protocol. The following characteristics of IsoMAC are analyzed in the subsequent sections.

- The upper limits of the approach in terms of number of real-time traffic flows for different update times
- The efficiency of the new protocol, especially when transmitting cyclic frames with a small payload, as compared to HCCA
- The temporal characteristics of the new proposal when several nodes exchange real-time traffic flows as well as best effort traffic

Even though the protocol mechanisms to ensure reliability are very important, they require a dynamic scheduling of the traffic to be retransmitted and an admission control which considers additional resources for retransmissions. Hence, they will be further discussed and investigated in Ch. 6.

5.4.1 Upper Bounds of the IsoMAC Approach

Usually, the IEEE 802.11 standard does not specify any limitations regarding the number of associated nodes to one access point (AP) for the DCF or EDCA medium access. In these modes it is also not able to provide any real-time guarantees. Since the proposed isochronous medium access control is designed to support real-time class 2, it is very important to determine the possibilities and limitations of the proposed mechanism. One of the most important properties from the application perspective, is the number of associated nodes to one AP and the corresponding possible update time. In this subsection, the boundary will be defined by using analytical calculations and assuming realistic conditions.

Scenario

Due to the specified protocol data units (PDU), the timing, used modulation schemes, etc. of IEEE 802.11, certain boundaries exist in terms of maximum number of nodes for a given update time and payloads. Hence, the provided analytical results for an example scenario, shown in Fig. 5.8, in order to provide upper bounds for the medium access mechanism.

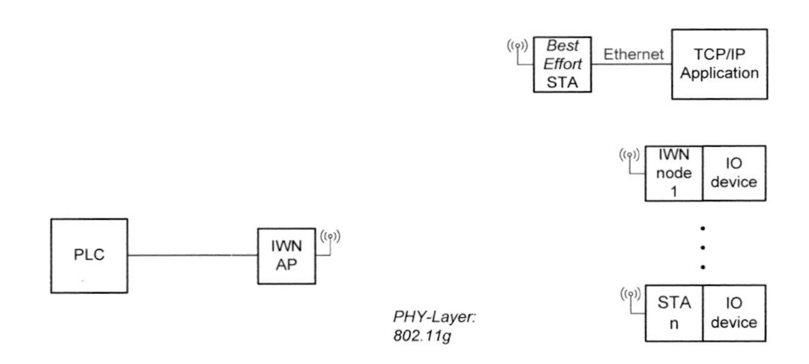

Fig. 5.8: Example scenario for the upper bounds estimation

The example scenario consists of a PLC as one traffic endpoint, n wireless nodes and an AP. The automation system is configured to work with typical update times for industrial applications of real-time class 2. The AP is the central element of the wireless network and in charge of coordinating all transmissions, i. e., it coordinates the medium access during the scheduled phase. The wireless data exchange between PLC and IO devices is realized using the physical (PHY) layer specified in the 802.11g amendment. It is assumed that best effort traffic is always transmitted with the highest possible data rate of 54 Mbps. Concerning bandwidth for retransmissions three different cases are considered and shown in Fig. 5.9.

(a) Ideal channel No bit errors occur during transmission, i. e., retransmissions are not needed
(b) Error-prone channel Occurence of bit errors, retransmissions are handled during the contention phase
(c) Error-prone channel Occurence of bit errors, additional bandwidth is reserved for retransmissions

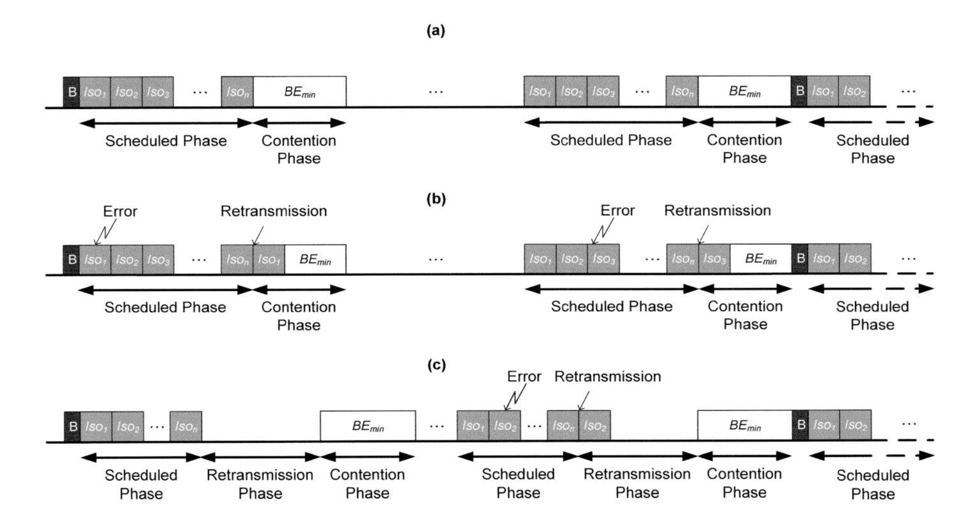

Fig. 5.9: Considered retransmission policies

Upper Bounds

The format of a 802.11g PLCP protocol data unit (PPDU) is depicted in Fig. 5.10. The MAC protocol data unit (MPDU) is first passed to the physical layer, which adds the physical layer convergence procedure (PLCP) header and the PLCP preamble. The PLCP header is always transferred with the most robust modulation scheme at a data rate of 6 Mbps and a coding rate of 1/2, which implies that every information bit is converted into two coded bits. The MPDU is transmitted with a variable data rate depending on the current signal quality. For the presented calculations data rates of 12 Mbps, 36 Mbps and 54 Mbps were considered.

The parameters of the 802.11g PHY layer used for the calculations are shown in Tab. 5.1. As it can be seen the number of coded data bits per OFDM-symbol varies, due to different modulation schemes. This value is directly related to the transmission time of a data frame. The duration for the PLCP preamble and the PLCP header remain constant.

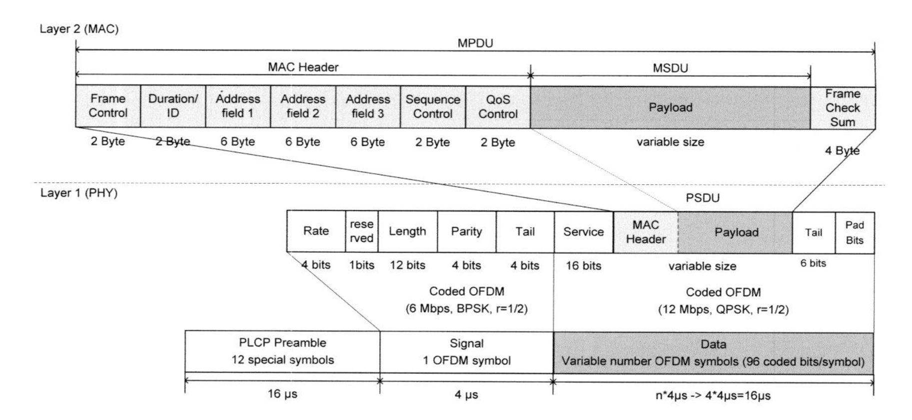

Fig. 5.10: Frame structure of an OFDM PPDU for 802.11g

Table 5.1: Parameters for the 802.11g physical layer

Abbreviation	Description	802.11g
t_{DIFS}	DCF interframe space	28 µs
t_{PIFS}	PCF interframe space	19 µs
t_{SIFS}	Short interframe space	10 µs
Slot time	Duration of one time slot	9 µs
$t_{PLCP\_preamble}$	PLCP preamble duration	16 µs
t_{PLCP_header}	PLCP header duration	4 µs
$Service$	Service field of PLCP header	16 bits
$Tail$	Tail bits	6 bits
r	Coding rate	1/2 (12 Mbps) 3/4 (36 Mbps) 3/4 (54 Mbps)
N_{CBPS}	Coded bits per symbol	96 bits (12 Mbps) 192 bits (36 Mbps) 288 bits (54 Mbps)
t_{OFDM_Symbol}	Duration of an OFDM symbol	4 µs
l_{MPDU}	MPDU Size (MAC header+Payload)	400 bits
l_{ACK}	MPDU Size (Acknowledgment)	112 bits

The duration of a data frame or an acknowledgment frame can be computed using (5.2). It depends on l_{MPDU} which is the size of the MPDU. For the presented results it was set to contain 50 bytes or 400 bits for a data frame, and 14 bytes or 112 bits for an ACK frame.

$$t_f(l_{MPDU}, N_{CBPS}) = t_{PLCP\_preamble} + t_{PLCP\_header} + \\ + \left(\frac{(l_{MPDU} + Service + Tail) \cdot \frac{1}{r}}{N_{CBPS}} \right) \cdot t_{OFDM\_Symbol} \tag{5.2}$$

Using Eq. (5.3) the total duration of a data frame exchange sequence t_{seq} in the scheduled phase can be calculated, whereas Eq. (5.4) is used to calculate a best-effort data frame exchange with the maximal possible payload of 2312 bytes.

$$t_{seq} = 2 \cdot t_{SIFS} + t_f + t_{ACK} \tag{5.3}$$

$$t_{BE\_seq} = t_{DIFS} + t_{SIFS} + t_{f_{\max}} + t_{ACK} \tag{5.4}$$

It is assumed that the network update time (NUT) of the IWN has to be less or equal to the required update time of the automation system, and that all real-time stations have the same temporal requirements. Further, the remaining standard stations shall be at least able to transmit one best-effort frame per NUT. Thus, the maximal number of traffic streams TS_{max} can be calculated with Eq. (5.5), if no retransmissions occur or if the contention phase is used for retransmissions (cf. Fig. 5.9 (a) and (b)).

$$TS_{\max} = \left\lfloor \frac{NUT - t_{BE\_seq}}{t_{seq}} \right\rfloor \tag{5.5}$$

When a separate retransmission phase is introduced, which has the same duration as the scheduled phase (cf. Fig. 5.9 (c)), TS_{max} is computed with Eq. (5.6).

$$TS_{\max} = \left\lfloor \frac{NUT - t_{BE\_seq}}{2 \cdot t_{seq}} \right\rfloor \tag{5.6}$$

The results of the calculations are shown in Fig. 5.11 for different network update times ranging from 1 ms...8 ms and data rates of 12 Mbps, 36 Mbps and 54 Mbps. For the highest requirement, i.e. a NUT of 1 ms, and the lowest data rate of 12 Mbps, it is possible to have at most 5 traffic streams. If a data rate of 54 Mbps is used, TS_{max} is increased to 7.

5.4.2 Efficiency of IsoMAC

The inefficiency of existing polling based mechanism is discussed in Sect. 5.2. Existing differences are depicted in showing the frame exchange sequence of both mechanisms. In order to evaluate the advantages of the proposed IsoMAC, the relative transmission time for payload frames is simulated. It is defined as the ratio between the frame transmission duration of the proposed IsoMAC (t_{IsoMAC}) and HCCA (t_{HCCA}) including the protocol overhead. This metric shows the efficiency of both protocols in a direct relation with respect to different payload sizes.

For both protocols, HCCA and IsoMAC, the frame duration is computed for a fixed data rate of 12 Mbps with Eq. (5.2), where l_{MPDU} includes the MAC header

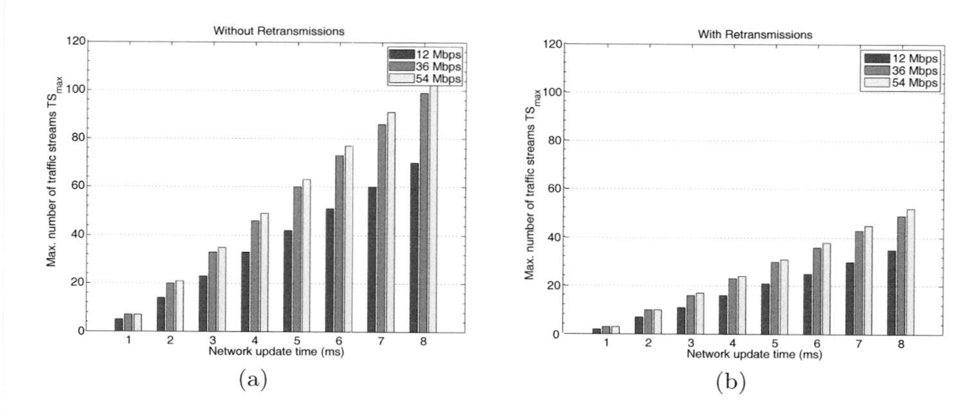

Fig. 5.11: Upper bounds for traffic streams using different retransmission policies

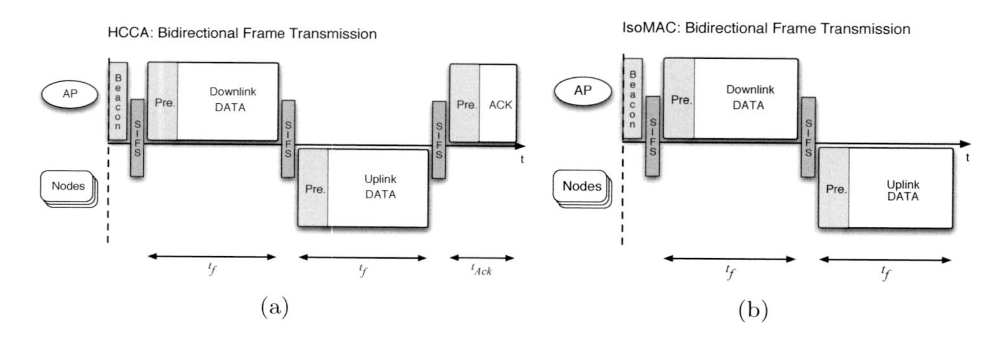

Fig. 5.12: Time for bidirectional frame transmissions of HCCA and IsoMAC

size of 30 bytes and the payload size, and N_{CBPS} = 96 bits for the data rate of 12 Mbps.

The frame exchange duration for HCCA having bidirectional traffic (uplink/downlink) is calculated using Eq. (5.7), with t_{SIFS} = 10 µs, and t_{ACK} being the duration of an acknowledgement frame. There are no separate polling frames, since they can be piggybacked with the data frames for the bidirectional case. The duration of an IsoMAC sequence is calculated with Eq. (5.8).

$$t_{HCCA} = t_f + t_{SIFS} + t_f + t_{SIFS} + t_{ACK} \tag{5.7}$$

$$t_{IsoMAC} = t_f + t_{SIFS} + t_f \tag{5.8}$$

$$G_{Iso} = \frac{t_{HCCA}}{t_{IsoMAC}} \tag{5.9}$$

To further investigate the IsoMAC efficiency, the frame transmission time including all protocol overheads was simulated with different payload sizes (varying from 0 - 1500 bytes) for HCCA and IsoMAC. The relative transmission time expressed as IsoMAC gain G_{Iso} and calculated with Eq. (5.9) is shown in Fig. 5.13, i. e., if $G_{Iso} > 1$ IsoMAC is superior as compared to HCCA.

The results show evidently that IsoMAC is superior for uplink traffic and for uplink/downlink traffic. Especially, in cases with typical industrial payload sizes ($\leq$ 100 bytes), a huge gap between HCCA and IsoMAC transmission times can be identified. For instance, considering a payload of 50 bytes, the IsoMAC uplink traffic consumes 55% less time than HCCA. It decreases for the uplink/downlink case, but still consumes more than 20% less time. Except when having downlink traffic only, the frame transmission times are identical.

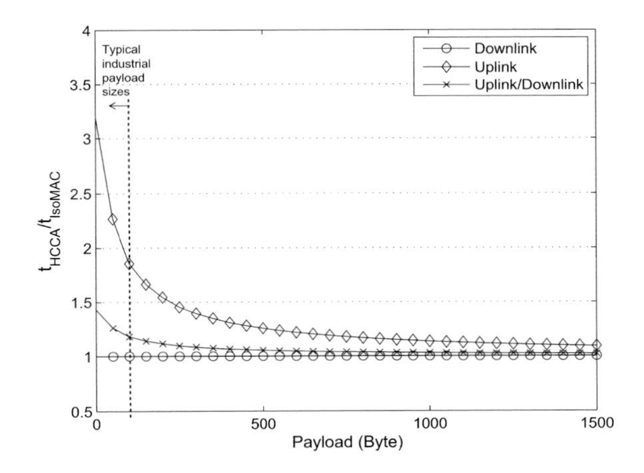

Fig. 5.13: Relative transmission time

5.4.3 Temporal Characteristics

To evaluate the temporal characteristics of the proposed medium access control, two different simulation experiments have been carried out, both consisting of two different scenarios. The first experiment has the objective to show that additional best effort traffic has no influence on the real-time traffic flows. The second experiment investigates the influence of an increasing number of nodes to the medium access in terms of the update time T_{IAT} of real-time traffic flows. First, the simulation scenarios are introduced, followed by the obtained results and their discussion.

Scenarios

In the first experiment, one AP with six associated nodes is used. Each node has one uplink traffic flow and one downlink traffic flow with a payload of 64 bytes.

In addition to this, 6 additional nodes exchange BE traffic with a throughput of 600 kbps per node, i. e., each node is approx. transmitting a frame with a payload of 1500 bytes and a rate of 50 s ^{1} representing a very bad case. The NUT T_{NUT} was set to 20.48 ms and the real-time traffic is transmitted with a data rate of 12 Mbps, and it is assumed to have an error free channel to be able to observe only the influences of the additional BE traffic. In the first simulation scenario the protection mechanism provided by the restricted phase is deactivated, whereas it is activated in the second simulation scenario.

In the second experiment, one AP with a varying number of nodes is used. The distance between transmitter and receiver ranges from 5 m up to 20 m and results in an average packet error rate λ_{PE} of 10% which is rather pessimistic. The communication is bi-directional, i. e., every node has one uplink flow with a payload of 64 bytes which is sent to the PLC connected via a RTE and one downlink flow with the same parameters sent from the PLC to the nodes. As in the previous scenario 6 additional nodes exchange BE traffic with a throughput of 600 kbps per node. The NUT T_{NUT} was set to 10 ms. The first simulation scenario is using the IWN simulation model (cf. Sect. 8.3). In the second scenario, the HCCA simulation model of [169] is used. The number of nodes participating in the scenarios and their application settings are shown in Table 5.2. The sample size for both scenarios is determined to be $N = 10000$ to obtain a confidence level of 95%.

Table 5.2: Application settings for the second experiment

No. of nodes	**10 ms**	**20 ms**	**40 ms**
12	2	4	6
24	4	8	12

Simulation Results

The results of the first experiment are provided in terms of the cyclic transmission time T_{NUT} of the beacon. The results shown in Fig. 5.14 are obtained from the first scenario where the protection mechanism of the restricted phase is deactivated. The deactivation leads to a foreshortening of the Scheduled Phase (SP), because ongoing BE transmissions occupy the medium and the beacon can not be transmitted at its scheduled point. Depending on the transmission duration of the BE frame including ACK, the beacon transmission will be delayed which negatively influences the whole SP. Using Eq. (5.2) and Eq. (5.4) this transmission time can be estimated to be 1.1 ms for the given setup. The mean value of T_{NUT} is 20.4855 ms with a standard deviation of 0.17985 ms, a minimum of 19.485 ms and a maximum of 21.487 ms.

The results for the second scenario are shown in Fig. 5.15.In this scenario the restricted phase is active and prevents the transmission of BE traffic towards the end of a NUT. Hence, the variation of the beacon transmission is dramatically reduced,

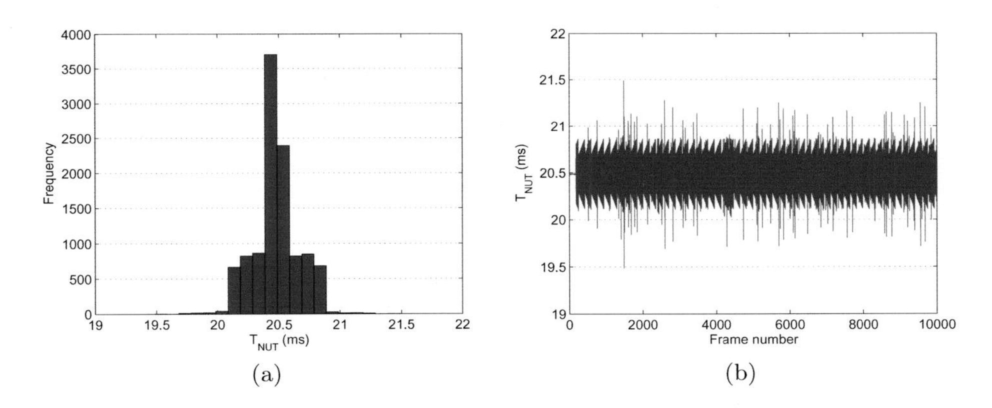

Fig. 5.14: Variation of the beacon transmission time T_{NUT} without restricted phase

and the obtained results a far better. The mean value of T_{NUT} is 20.4853 ms and almost equal to the previous scenario. However, the standard deviation is 0.00220 ms which is only 1% of the first scenario. The minimum is determined to be 20.478 ms and the maximum is 20.49 ms with a total difference of 12 µs.

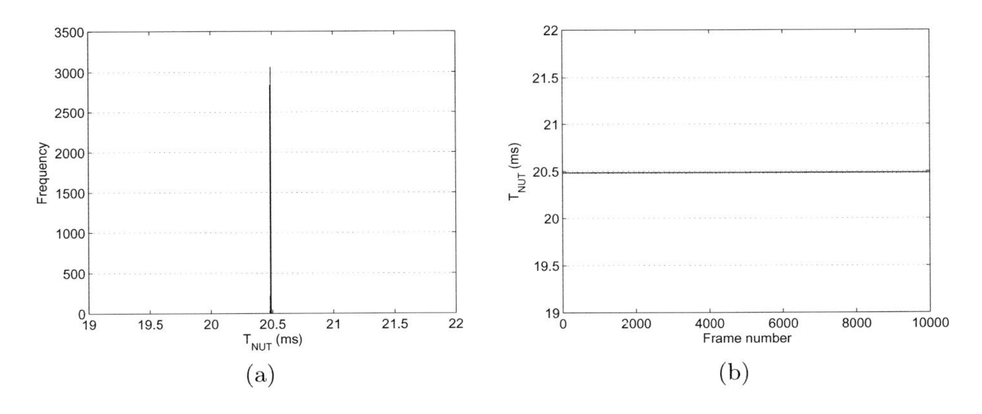

Fig. 5.15: Variation of the beacon transmission time T_{NUT} with restricted phase

The results of the second scenario are presented as a comparison of the new isochronous approach and the HCCA mechanism with respect to the achievable update time T_{IAT} for the traffic flows and its corresponding communication jitter. In Fig. 5.16 the update time distribution of all traffic flows for the scenario with 12 nodes is shown. For instance, when using HCCA the 10 ms flows have a mean value of 10.1209 ms and a standard deviation of 11.2603 ms. Whereas the IsoMAC approach achieves a mean value of 10.0556 ms and a standard deviation of 0.9985 ms which is mainly caused by retransmissions.

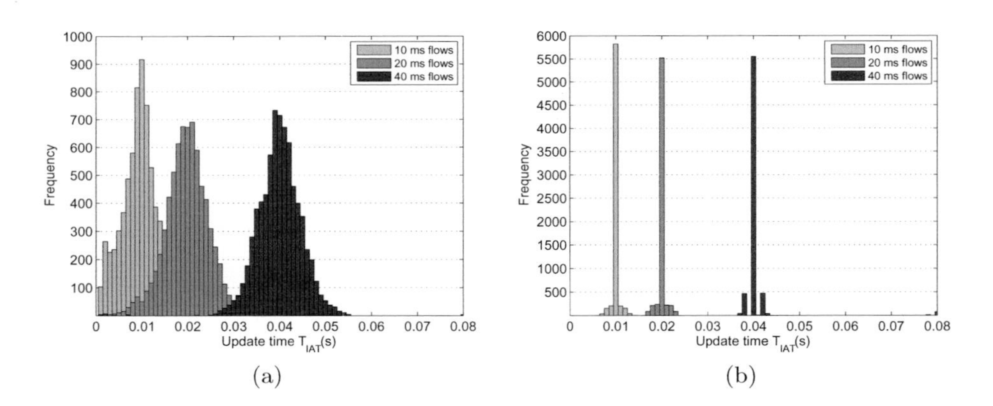

Fig. 5.16: Update time T_{IAT} for the HCCA and IsoMAC scenario with 12 nodes

The update time distribution for 24 nodes is depicted in Fig. 5.17. Although the update times and their deviation are slightly increased for the new approach, due to the higher number of flows, it still outperforms HCCA. For every traffic flow (10 ms, 20 ms, and 40 ms) the update time standard deviation of the new mechanism is smaller than for the HCCA case. Even though the 10 ms flow might indicate a slightly widened characteristic, it has only a standard deviation of 1.5285 ms as compared to 4.8355 ms for HCCA.

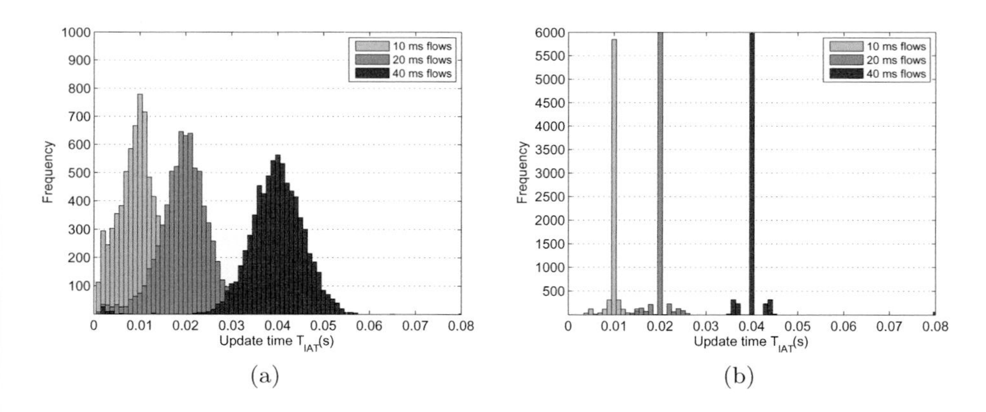

Fig. 5.17: Update time T_{IAT} for the HCCA and IsoMAC scenario with 24 nodes

It can be observed in both scenarios that the behaviour is almost not influenced by the number of nodes and real-time traffic flows. This would have been expected for both protocols due to their basic principle of coordinated access to the medium. However, the HCCA shows a very high deviation of the time of transmission, even though the QoS guarantees in terms of throughput and service interval are provided as initially allocated. The reason for this is the strong orientation of the 802.11 protocol towards throughput optimization. Hence, it is only considered that the traffic flow with its maximum service interval, but the instant of time for the transmission is not relevant.

5.5 Concluding Remarks

In this chapter a new approach for a deterministic Medium Access Control (MAC) is introduced which addresses the identified challenges of an efficient transmission of real-time traffic with small payloads, the reliable wireless transmission, and the integration of standard nodes without any influence on the real-time traffic.

The previous analysis results have shown that the proposed MAC protocol is able to provide an efficient mechanism to distribute the required communication schedule as well as to deal with error recovery. A protection mechanism assures that the beacon frame and the associated scheduled phase always start as intended with a very small jitter of approx. 10 µs. This is a very crucial property of the protocol, because interferences of non-real-time nodes must be avoided, at the same time the transmission of BE traffic must be possible with a sufficient throughput capacity.

The evaluation results clearly show that our mechanism is superior to existing polling-based MAC protocols as specified in 802.11, especially in terms of communication jitter of the update time. Whenever the channel quality decreases and retransmissions occur, the efficient and flexible retransmission handling is able to retransmit the erroneous frames within the same NUT interval. However, this depends on the application, if retransmissions are not desired.

6

Resource Allocation

In this chapter an approach for the resource allocation component for the IWN is proposed. The resource allocation is responsible for allocating the required resources to nodes in the network. It considers currently admitted traffic flows of the AP and the available capacity of the channel. It also guarantees that the application requirements can be met by all admitted flows.

Different problems must be considered in order to be able to satisfy the requirements of the application. An unpredictable behaviour of the wireless network due to an uncontrolled increased utilization and the resulting overload situation in the network must be avoided. The component consists of three main building blocks, the admission control, the scheduler, and the adaptive retransmission handling. After an analysis of related work in the field of admission control and scheduling for TDMA-based wireless networks, existing industrial traffic types are defined. Afterwards the main building blocks are described and analyzed.

6.1 Admission Control and Scheduling in TDMA-based Wireless Networks

Within this section the related work for admission control and scheduling approaches for TDMA-based wireless networks is discussed and analyzed with respect to the given requirements and the aforementioned challenges imposed by the application. In addition to this, typical traffic types in industrial systems are defined.

6.1.1 Related Work

In the context of scheduling and admission control for IEEE 802.11 analyze the stand and propose new algorithm to improve the original proposal of the HCCA. Both traffic types, Constant Bit Rate (CBR) as well as Variable Bit Rate (VBR) traffic are considered for voice and video applications. However, for this work only CBR traffic is important and discussed.

In [16] the inefficiency of the HCCA is considered under the assumption that the parameters have not been modified with respect to the utilized application, i. e., configuration issues with the medium access mechanisms are investigated. However, the main focus is put on client/server file transfer applications and multimedia traffic.

Hence, a different scheduling mechanism are considered by Cicconetti et al. in [24]. Cicconetti et al. propose an algorithm to reduce the polling overhead of HCCA for uplink transmission as much as possible by combining several TXOPs to a single one wherever possible. Further, the computational intensive calculations are performed offline, while the enforcement procedure is performed online. This approach is again optimized for throughput and multimedia applications.

Junior et al. propose in [82] a scheduler especially designed for industrial automation systems with the result of having a higher efficiency and reducing the polling overhead by using a token passing procedure.

A scheduling algorithm being compatible with link adaptation algorithm is proposed by Grilo et al. in [48] for the 802.11 hybrid coordination function. The algorithm bounds the time of medium usage for each station. The performance of the approach is compared to the original scheduling approach by using simulations.

Lim et al. [99] describe an algorithm for ensuring fair access of a granted TXOP by combining them into a single *First In, First Out (FIFO)* queue to different streams at a station. For the reception of a poll, this mechanism allows for a minimized MAC processing. Moreover, the delivery of QoS sensitive traffic is optimized by a novel scheduler.

The problem of variable bit rate traffic is addressed by a new approach of Ruscelli et al. [148]. They propose a new scheduler using the negotiation procedure of HCCA for a minimum bandwidth. For traffic streams which need more bandwidth than the negotiated one, a redirection of the additional bandwidth to EDCA is done. Improvements regarding the throughput, access delay and the queue length could be achieved.

An adaptive TXOP allocation is proposed by Arora et al. [2] for a simple Hybrid Coordination Function scheduler which takes the QoS requirements of admitted flows into account. It takes the channel and traffic conditions into account and complies with the rate adaptation mechanisms. The algorithm is validated with simulation studies showing the improved performance of it.

Toscano et al. [160] propose a scheduling framework based on EDF which can be deployed in cellular networks where the traffic flow is traversing more than one wireless link. This has been formulated by precedence constraints applied during the schedulability analysis.

Discussion Within the analysed works, the integration into a decentralized TDMA-based medium access control is not considered, except for the last one. Usually, it is assumed that the wireless network either uses the centralized polling-based HCCA or even the EDCA based on a priority assignment to traffic flows. Both topics are deeply investigated, but with a strong focus on multimedia related scenarios. Furthermore, a consideration of the wireless channel condition at a given time has only been done by one of the presented works, especially retransmission handling is excluded. However, this must be considered to guarantee that there are no overload conditions due to retransmissions and hence allow a reliable transmission whenever it is needed. Most of the works focus on multimedia scenarios with the main goal to optimize throughput and less considering the temporal properties of traffic flows such as the latency jitter.

6.1.2 Industrial Traffic Types

According to [7], traffic in industrial networks can be categorized into 3 different types: (i) Real-time Cyclic (RTC), (ii) Real-time Acyclic (RTA) and (iii) BE data. RTC data occurs typically in control applications. RTA data occurs for instance, if alarm signals are issued. The BE data type is used during file transfers or configuration traffic. Whereas the third traffic type has typically no temporal constraints, both other traffic types (i) and (ii) have to meet temporal requirements, that depend on the process, i. e., the data needs to be transmitted within a specific time frame, to avoid asynchronously operating process modules, which might cause a degraded product quality.

Real-time Traffic

The real-time traffic comprises the cyclic transfer of process data for closed loop control referred to as RTC and the acyclic transfer of events such as alarms referred to as RTA. Both real-time traffic types can be distinguished as follows.

Real-time Cyclic traffic flows transfer process data in predefined periodic cycles from the PLC to the IOD and vice versa. Since an NCS is also based on a periodic sampling strategy, this type of traffic is used for closed loop control. Real-time cyclic traffic in RTEs has the following properties:

- RTC traffic flows are always delay sensitive and must be transmitted before an upper latency bound, referred to as deadline d_P which depends on the requirements of the application
- RTC traffic flows are sent in a cyclic pattern characterised by the send period T_P
- RTC traffic flows consist of a payload l with a constant size, the cycle time and the constant size result in CBR traffic
- The omission degree is monitored by the application, as long as no new process data value is received, the old value is used until a given omission degree OD is exceeded. This must be avoided, because it will result in a failure of the system
- When following the producer-consumer communication model, RTC traffic flows are transferred unacknowledged, but their arrival is usually monitored by an application watchdog
- In case RTC flows must be buffered, because the transmission is impossible, the RTC data becomes obsolete and should be dropped with the next update cycle time

Real-time Acyclic traffic is used to indicate events from automation devices to the PLC. Events may be system defined events such as plugging and unplugging of modules and user-defined events like an increased temperature. For example, acyclic real-time communication has the following properties in Profinet systems, but is similar in other systems.

- RTA must be acknowledged by the PLC on the application layer
- RTA data is delay sensitive, but in the range of several hundred milliseconds should be forwarded with high priority
- RTA has defined omission degree, too many erroneous consecutive frames result in a stop of the whole system

Even though the scheduling mechanism allows to schedule this type of real-time traffic, it is not further considered in this work, because it is usually less important and subordinated to the cyclic real-time traffic [42].

Non Real-time Traffic

The non real-time traffic is used to set up and to monitor the context for real-time traffic flows and to configure IOD. The real-time traffic flows will be started after this context has been established. The following non real-time services are commonly used for this purpose in automation:

- Discovery of automation devices and basic configuration services are used to find devices in the network and assign or resolve naming and addressing information. They are based on DCP, LLDP, ICMP, DHCP and DNS.
- Context management services establish application relations between PLC and IOD. The successful establishment of this context is the prerequisite for the transfer of real-time traffic flows. They are based on RPC (connectionless).
- Configuration and diagnosis services are used to set up or read out device configuration parameters. They are based on SNMP, IGMP, RPC (connectionless), http, and ftp.

In addition to this, non real-time traffic is required for monitoring and maintenance purposes. For example, maintenance personnel accessing the webserver of certain devices to obtain their current status or other monitoring information as in the first application category, described in Sect. 1.1. Video streaming for surveillance and visual inspection of the plant becomes also increasingly important, especially when remote locations are to be monitored, as introduced in the second application category in Sect. 1.1.

6.2 Admission Control

Whenever a wireless real-time node is added to the wireless network, e. g., a flexible module is added to the manufacturing system, it has to perform an admission procedure in order to specify its needed resources. Based on a schedulability test, the admission control module decides whether sufficient capacity is available to guarantee the needed resources including reserves for adaptive retransmission handling or not. The whole resource allocation module is shown in Fig. 6.1.

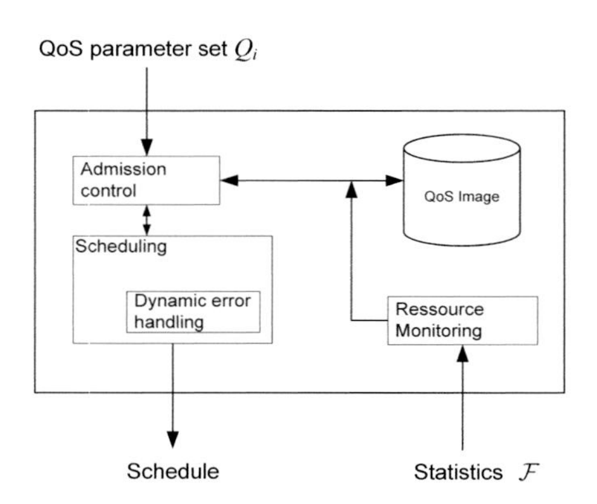

Fig. 6.1: Resource allocation component overview

The admission control itself is based on priorities and allows the rejection of flows in case a flow with a higher priority enters the system, and there is not enough capacity to accept this new flow. The admission control maintains a QoS image of the system which is an exact image of all admitted real-time traffic flows. Furthermore, the admission control is interacting with the traffic inspection unit (resource monitoring) which continuously observes the application layer of the connected RTE system. During the admission control procedure, as defined in Sect. 6.2, the admission control gets the QoS parameter set Q_i which contains all relevant information. Q_i will be either provided by the engineering or by the traffic inspection. Based on this information a pessimistic schedulability test is performed in order to provide reserves for retransmissions whenever they are needed. If the schedulability test is successful, the new flow will be accepted and the updated schedule will be provided to the deterministic medium access control module.

6.2.1 Retransmission Behaviour in Industrial Environments

In order to gain a better understanding of the nature of transmission errors in industrial environments, the occurrences of consecutive retransmissions in a real industrial manufacturing environment are investigated in [31]. The approach and the main results are briefly described, since they are part of the basis for the remaining chapter.

Testbed and Scenario

The industrial environment is the same as described in the evaluation (cf. Sect. 8.4). The measurements are carried out during normal working hours. All scenarios consists of an AP and a node, both are mounted at a height of approximately 1,5 m from ground, the transmit power is set to 20 dBm. The Anritsu data analyzer (model MD1230B) is used as traffic generator, and the wireless traffic is captured and recorded using a wireless monitor which is carefully collocated next to the node. Altogether four scenarios are setup which are configured as follows:

- WLAN technology: IEEE 802.11g
- Payload size: 64, 128, 256 and 1518 bytes
- NUT: 10 ms
- Data rate (modulation and coding rate): 12 Mbps, (QPSK, 1/2), 54 Mbps, (64-QAM, 3/4)
- Wireless link: 15 m Line-of-sight (LOS) and 20 m Non Line-of-sight (NLOS)

The parameters are selected according to typical factory automation systems.

Consecutive Retransmissions

After the experiment, the occurrences of consecutive retransmissions are extracted from the trace files. They are shown in Fig. 6.2 for the two scenarios with a data rate of 54 Mbps and different payload sizes.

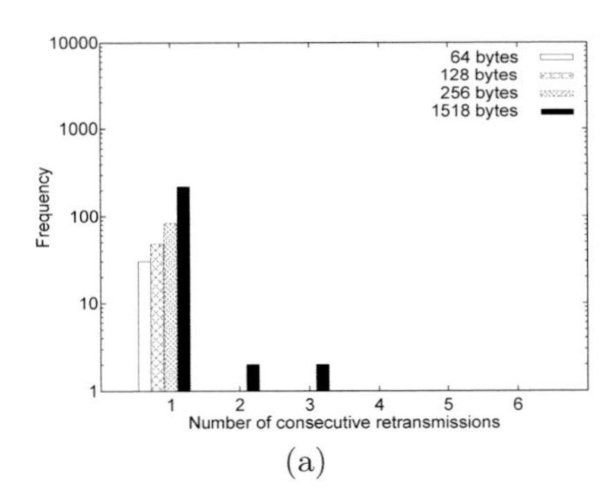

(a)

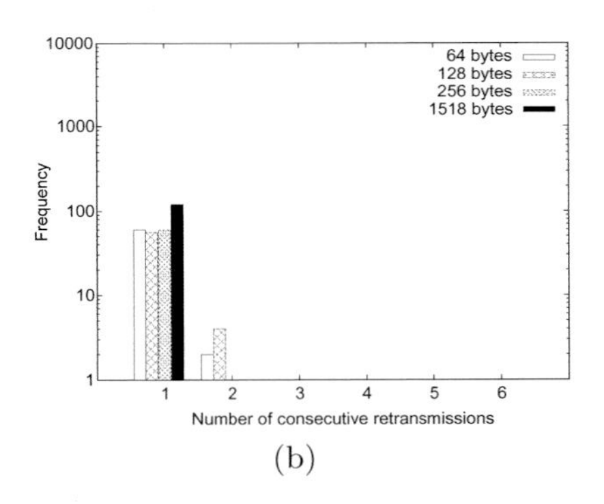

(b)

Fig. 6.2: Consecutive retransmissions in an industrial environment for 54 Mbps/LOS (a), and 54 Mbps/NLOS (b)

Consecutive retransmission occurrences at a data rate of 54 Mbps with LOS and NLOS link as shown in Figs. 6.2(a) and 6.2(b) respectively. In both diagrams a trend of greater number of retransmissions as the frame size increases can be observed. Moreover, the distribution of retransmissions for both links is concentrated on one or, seldom, two retransmissions.

For a data rate of 12 Mbps the consecutive retransmissions show a similar characteristic, but they are further decreased due to the more robust modulation and the higher FEC coding rate as shown in Fig. 6.3. Again, smaller frames are more

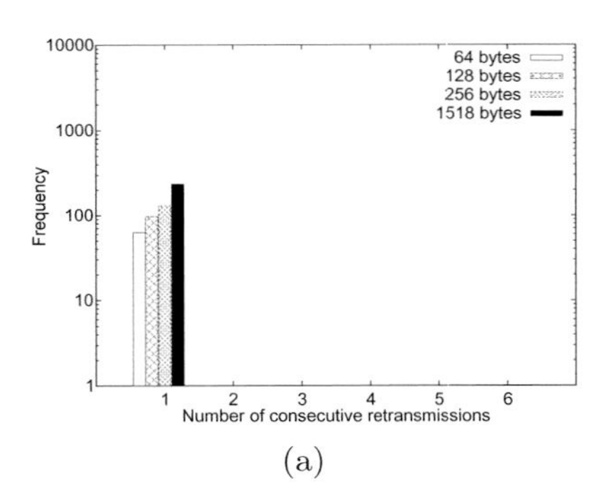

(a)

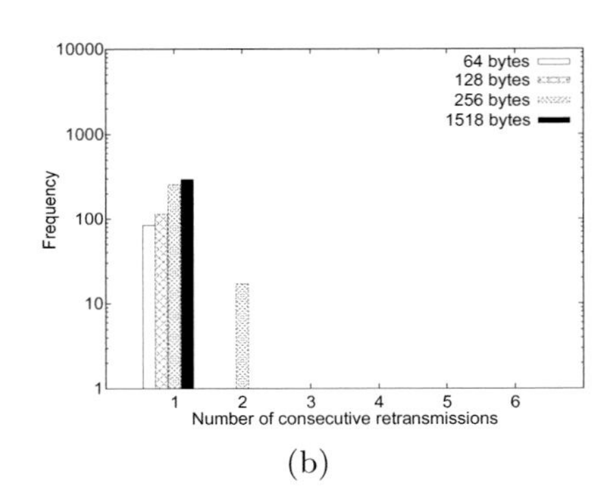

(b)

Fig. 6.3: Consecutive retransmissions in an industrial environment for 12 Mbps/LOS (a), and 12 Mbps/NLOS(b)

likely to be transmitted without experiencing errors. Especially, industrial traffic flows with small payload sizes will be less affected.

The obtained results of this investigation are particularly relevant for error recovery resource estimation in the proposed IWN. These results provide a guideline of how often retransmission efforts for a traffic flow could be needed and how many retransmissions will be required to successfully deliver a frame that belongs to that traffic flow.

6.2.2 Cross-layer Traffic Inspection

The cross-layer traffic inspection unit has two major task. First, it should derive a QoS parameter set during the RTE connection establishment, if possible. Second, the monitoring of the real-time traffic flow on the application layer, to provide direct input to the dynamic retransmission unit described in Sect. 6.4.

The engineering procedure of RTEs, as described in Sect. 2.1.1, leads to a static network configuration. Therefore the scheduling for the wireless network has to be done once during the start-up phase of the system. All relevant nodes of the hybrid system, including active and inactive wireless nodes, are considered. Therefore, the information about QoS requirements of the application are already available in the engineering system. For instance, when listening to the initial frame exchange of devices during the connection establishment according to the protocol (similar to establishing a session), it is possible to derive these requirements and use them for an admission procedure in the next step. This method requires perfect knowledge and interpretation of the relevant RTE protocol.

Up to now, RTE protocols do not offer mechanisms to provide information to other networking infrastructure components, such as wireless systems, about their specific QoS requirements of the application and whether they are fulfilled. However, this information is very useful either to obtain a valid QoS parameter set for the admission procedure within the wireless network or for monitoring purposes on the application layer as described in

Existing methods for traffic flow identification and monitoring, such as in [14, 15], are based on frame statistics. This requires the observation of the traffic flow for a sufficient time interval. It would not be suitable for generating a QoSparameter set for the admission procedure. The method proposed in this work is based on cross-layer packet inspection. There are two main steps: RTE protocol identification and Traffic Flow (TF) signature derivation.

The packet is inspected to identify what protocol it belongs to. The TF signature is derived based on the protocol that has been identified. For both steps, we use a set of parameters that are unique to the specific RTE protocol, and that can be contained on one or more layers (link, network, transport, application, etc.). The signature is a unique set of values that are identical in every packet that belongs to a particular TF. Note that, in practice, every packet is parsed only once and that protocol identification and signature derivation can be done in a single step. Once a TF has been identified, all related statistics can be derived: update time, jitter, omission degree, etc.

6.2.3 Resource Reservation and Admission Control

The allocation of real-time traffic flows to transmission slots is performed by a scheduler, which resides in the IWN AP. The aim of the scheduler is to produce a feasible schedule, i. e., a schedule meeting the deadlines of all real-time traffic flows. It is not possible to guarantee that all traffic flows will be correctly received within their deadline, because of the unpredictable wireless medium. However, in a feasible schedule, it is guaranteed that all the real-time flows are transmitted before their deadlines expire. If a real-time message is lost due to channel errors, it will be retransmitted, using the retransmission mechanism described in Sect. 5.3.

According to IsoMAC, retransmissions of a scheduled flow always occurs within the same superframe as the one in which the message is scheduled. Therefore, to ensure that flow retransmissions will also occur before their deadline, it is sufficient to configure flow deadlines d_i that are multiples of the T_{NUT}.

As previously introduced, two different categories of real-time flows are considered, periodic traffic flows and aperiodic traffic flows. Flows of the first category are characterized by a totally predictable arrival pattern, with fixed period T_{P_i}, relative deadline d_i and communication time C_i, based on the payload l_i. As already mentioned, the vast amount of industrial real-time traffic is periodic [42]. Flows of the second category feature a fixed relative deadline D_i and maximum communication time C_i, while the arrival time is unpredictable, but very infrequent. Since the sending rate of the aperiodic flows is much lower than the system dynamics, at most one occurrence in the schedule is assumed at this point to provide real-time guarantees to this kind of traffic.

Both kinds of real-time flows must be admitted before the relevant transmissions can start. Once admitted, they are guaranteed to transmit their message before their deadline expires. From the application point of view, there is no difference between the admission control of periodic and aperiodic flows. In both cases, an *admission request* frame must be sent to the AP, which contains all the information needed by the admission control and the scheduler to test the feasibility of the flow set. Moreover, in both cases the outcome of the schedulability test is notified to the requesting node through another special frame, called *admission response*. However, periodic and aperiodic flows must be handled by the proposed scheduler in different ways.

While periodic flows have timeslots that are directly assigned to each of them, aperiodic flows share common timeslots, which are allocated to specific flows whenever needed. Here, flows are dynamically scheduled and data transmission is guaranteed to occur before flow deadlines thanks to the bandwidth reservation mechanism based on aperiodic servers. In this way, bandwidth is allocated in a more efficient way than in pure TDMA systems where every flow has its time slot.

This proposal for the scheduler is based on the ^{flex}WARE approach [163] and the work of Toscano et al. in [160, 169]. The scheduling of periodic flows is discussed in Sect. 6.3.1. Even though it is of minor importance for this work, the scheduling of one-shot flows is discussed in Sect. 6.3.2. Before the actual scheduling procedure is described, the structure of a transmission schedule which must fit to IsoMAC is introduced.

6.3 Scheduler

The timeslots for transmissions are assigned by the scheduling module of the IWN AP to periodic traffic flows using a deadline-driven scheduling algorithm. First it deals with periodic flows by calculating the schedule for the length of the hyperperiod, which is defined as the least common multiple of the periods of all traffic flows. It is called major cycle T_{MC}.

The major cycle T_{MC} consists of an integer number of T_{NUT}. In Fig. 6.4 the mapping of a minor cycle to the IsoMAC communication cycle with period T_{NUT} is shown.

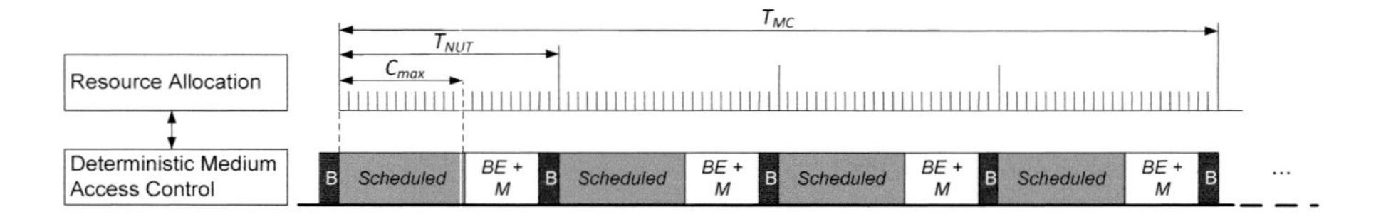

Fig. 6.4: Mapping between IsoMAC superframe and transmission schedules

6.3.1 Scheduling and Admission Control of Periodic Flows

The scheduling approach allocates the bandwidth per traffic flow. This means that each periodic real-time flow has assigned timeslots for its exclusive use. This approach provides more flexibility than the classical solution of assigning timeslots to stations, as it is easy to support nodes with mixed traffic requirements. Moreover, the schedulability of such an approach is improved.

The scheduling algorithm for periodic flows is dynamic and deadline-driven, i. e., it assigns higher priorities to flows with smaller absolute deadlines, which is the relative deadline plus the arrival time of the flow. Therefore, the scheduling algorithm can be considered an extension of Earliest Deadline First (EDF), that has to meet some additional constraints not addressed by EDF, but fundamental in this context.

An extension of EDF is used, even though EDF is not optimal when the assumption of preemptive scheduling is released. However, it generally improves schedulability as compared to other scheduling algorithms [49]. It is shown in [81] that no optimal scheduling algorithm exists for a set of periodic and sporadic non-preemptive tasks.

Two constraints are imposed by the fact that the transmission schedule calculated by the AP must adhere to the IsoMAC superframe. First, a minimum number of timeslots must be reserved for the CP of IsoMAC, because the CP is used to transmit best effort traffic and management traffic. The scheduler handles this constraint by storing the workload of each minor cycle in a QoS image, i. e., the amount of slots used by the currently allocated flows. It is ensured that it never exceeds the maximum duration C_{max}. It depends on T_{NUT} and on the retransmission policy as discussed in Sect. 6.4.2.

Second, all downlink slots, i. e., slots containing messages sent from AP to node, have to be scheduled before the uplink slots, i. e., slots containing messages sent from the nodes to the AP. This constraint is handled by the scheduler by storing two different flow queues for each minor cycle, i. e., one queue for downlink and one for uplink flows. To ensure that a newly inserted downlink flow does not cause a deadline miss, the scheduler also stores the maximum slack time of the uplink flows allocated in each minor cycle. The slack time is defined as the temporal difference between the deadline and the finishing time, $d_i - f_i$ of the communication cycle T_{NUT}. A new downlink flow can be inserted to the minor cycle only if twice its duration is smaller than the slack time of uplink flows already allocated in that minor cycle minus the required reserves for retransmissions.

The format of the output of the scheduler is similar and aligned to the IsoMAC beacon frame. However, it also contains additional information, which is used by the scheduler to include aperiodic flows into the actual schedule sent through IsoMAC beacon frames.

The transmission schedule of periodic flows is calculated by the run-time schedulability test during the admission procedure. The test executes the scheduling algorithm, which returns the transmission schedule as long as a feasible schedule exists. A defined error code is returned otherwise. As a result, the schedulability test is simply performed by checking the return value of the scheduling algorithm.

6.3.2 Scheduling and Admission Control of Aperiodic Flows

The aforementioned transmission schedule is calculated by the IWN AP and deals with periodic flows only. It does not include aperiodic flows directly. However, in each cycle T_{NUT} a defined number of timeslots is reserved. The timeslots can be dynamically allocated by the IWN AP to transmit real-time aperiodic flows. To reserve these slots, the scheduler inserts defined schedule information elements into the IsoMAC beacon, i. e., a special TFlowID field is used.

In this approach, aperiodic flows are allocated analogue to the Polling Server (PS) approach of real-time operating systems to schedule aperiodic tasks. For this reason, the number of timeslots allocated to aperiodic traffic is referred to as the *server capacity* (C_s). At the beginning of each minor cycle, i. e., before sending the IsoMAC beacon frame, the aperiodic scheduler is executed on the IWN AP. It checks whether there are pending aperiodic messages and allocate them into the reserved slots. As C_s is often much smaller than the sum of the C_is of all the aperiodic flows, the transmission time of all the pending aperiodic transmissions can exceed C_s. To provide a deterministic worst-case transmission time, aperiodic requests are served in increasing order of relative deadline and a timing analysis is performed when admitting new aperiodic requests.

The admission control of aperiodic flows verifies that the transmission time of each aperiodic flow is smaller than its deadline in the worst case, i. e., when all aperiodic flows are released at the same time. The computation of such a worst-case transmission time is based on the response time analysis of Polling Server provided in [12], which has been adapted for the specific case of this system. It should be noted that the worst-case transmission time of aperiodic flows depend on the server capacity. As a result, it is important that the admission control is able to find a suitable C_s value that makes the set of aperiodic flows schedulable without wasting too much bandwidth for other traffic. The approach followed by the scheduler consists of two steps. First, for each aperiodic flow, it finds the smallest server capacity that meets the schedulability condition. Second, it selects the actual C_s value as the minimum server capacity that fulfils the requirements of all the aperiodic flows.

6.4 Dynamic Scheduling of Retransmissions

In IEEE 802.11 infrastructure networks acknowledgements are used for unicast communication. If a station sending a unicast data frame does not receive an acknowledgement, it will interpret this as an unsuccessful transmission. The retransmission of the lost frame will be performed. The reasons for the sender of not getting an acknowledgment can have two reasons. First, the frame which contains data got lost or damaged, and the receiving node has nothing to acknowledge. Second, there can be a situation where the frame which contains data successfully reached its destination, but the acknowledgement got lost and never arrived at the sender. In both cases the node, which expects the acknowledgement after a defined *ACKTimeout* interval, starts to retransmit the frame again. This retransmission mechanism will be repeated a predefined number of times either until the frame is received or the maximum amount of retransmissions is reached. Therefore, retransmissions increase the communication reliability but, on the other hand, increase the latency of frames and their corresponding jitter [178] which should be avoided in an isochronous network.

Hence, an adaptive retransmission handling of traffic flows is proposed in this section. It is based on individual flows and their current status being continuously monitored by the traffic inspection unit. Depending on the application two different policies can be adaptively applied and will be introduced in this section.

6.4.1 Application Layer Considerations

Applications and the application layer of RTEs can tolerate omission failures up to a maximum omission degree OD_{max} which is usually 2, i.e., the connection will remain established provided that the given OD_{max} (started when the last data frame was successfully delivered) has not been exceeded. Therefore, it can be argued that there is no need to retransmit each frame at any price. Some frames can be lost, without influence on the industrial process (application). It is acceptable to loose some single frames, but it must be avoided that a number of consecutive frames, from one particular connection, are lost. An example which shows a burst of omission failures (not acceptable) and single omission failure (acceptable) is illustrated in Fig. 6.5(a) and in Fig. 6.5(b) respectively.

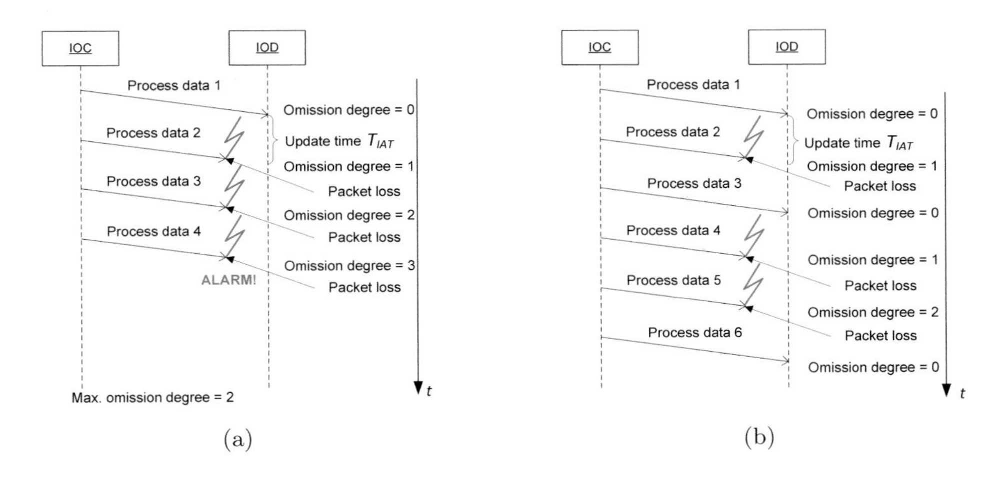

Fig. 6.5: Application behaviour when the maximum omission degree OD_{max} is exceeded

To exemplify the function of OD_{max}, Fig. 6.5(a) shows a scenario where three ($OD_{max} + 1$) consecutive frames from the PLC are lost, and an alarm is triggered by the IOD resulting in a connection abortion between IOC and IOD. This kind of situation must be avoided, because it means that the whole manufacturing process is stopped. In the second scenario (see Fig. 6.5(b)) some omission failures exist, but the maximum omission degree is not exceeded. In consequence it can be assumed that in this situation it is enough for the PLC to receive at least each third frame to avoid raising the connection alarm.

Moreover, depending on the amount of time needed for a given retransmission T_{retry}, the data in the retransmission queue is already outdated. This is the case when $T_{retry} \leq T_{NUT}$, because a more recent process data value has already arrived. The old process data is no longer relevant for the control loop and can be discarded by system.

6.4.2 Adaptive Retransmission Policies

Statically allocating resources for retransmissions during the CFP not only greatly reduces bandwidth efficiency, but also does not ultimately guarantee successful delivery of a frame. A scheme, in which CFP resources for retransmissions are granted beforehand, leads to bandwidth degradation for real-time traffic flows, because such resources may not be used at all during good channel conditions. The approach can also prove to be utterly ineffective during periods of time when the channel conditions are such that a higher number of retransmissions (greater than the available resources can allow) are required for successful frame delivery. In such case, all devoted bandwidth resources are only wasted.

Therefore, an adaptive retransmission approach is proposed. It allows for retransmissions to occur in a fashion that attempts to maximize reliability of traffic flows and does not impact other real-time flows. At the same time it can be also optimized with respect to the latency and latency jitter. Two different policies are described in the following paragraphs which allow both deployment scenarios. Common properties of the proposed mechanisms are listed below:

- No static resources will be allocated for retransmissions of erroneous frames within the CFP, the resources will be dynamically allocated if needed and appended to the CFP, i. e., there are no unused pre-allocated resources for error recovery
- Frames must be dropped when their deadline is expired, because only the latest process data is relevant
- According to Sect. 5.3, only downlink flows are acknowledged, either by an embedded or separate ACK frame

Policy 1

In policy 1, the isochronous property of the protocol is preserved, and it must be assured that the nodes are always in the valid coverage area with a bounded omission degree OD_{max}. In cases where this is no longer true, frames can be retransmitted if $OD \geq OD_{max}$, i. e., after OD_{max} frame losses. This will allow a graceful degradation of the application if necessary. This policy is relevant, if the application requires isochronous traffic with a minimal jitter. Retransmissions and the resulting jitter are only added if the system is in a state where the next frame must be transmitted to avoid stopping the whole system.

Policy 2

When using policy 2, erroneous frames will be always retransmitted with a defined and configurable number of attempts at the end of the CFP. The admission procedure will already consider the needed additional capacity for retransmission attempts when selecting the second policy. In this way, none of the other traffic flows, which might experience good channel conditions, is affected by the increasing efforts and resources devoted to recover from errors of specific flows. This policy allows a dynamic resource allocation, and the resources are only needed if retransmissions occur. Thus, the allocated bandwidth can be used by other best-effort and management traffic, if no retransmissions are possible. This policy leads to a higher reliability of the system and allows a deployment in more challenging wireless environments.

The selection of the policy is always a trade-off between the isochronous characteristics of the protocol and having a high reliability of frame transmissions even when the wireless channel quality degrades. The more important characteristic must be always determined depending on the relevant application and the environment.

6.4.3 Analysis of the Dynamic Resource Allocation

In order to assess the possibilities of the dynamic resource allocation presented in this section, several simulation scenarios are conducted with different objectives. The first scenario has the objective to determine the maximum number of flows, which can be successfully admitted using the presented admission control. The second scenario considers the dynamic retransmissions and investigates the influence of the packet error rate on the resulting packet loss rate for the application.

Number of flows

In the first scenario, the amount of traffic flows is considered depending on the chosen admission control. Two different approaches can be applied by the admission control module. First, it is assumed that each real-time traffic flow is only schedulable, if there is sufficient capacity to transmit the flow twice, i. e., a maximum utilization $U_{max} = 0.5$ is allowed. This is considered to be a pessimistic admission control. In the next scenario, this value is increased to $U_{max} = 0.66$. This utilization is considered to be the optimistic approach. The system is deployed with a varying network update time $T_{NUT} = 4\text{ms} \ldots 32\text{ms}$ and it is assumed that all new flows have identical periods with $T_{P_i} = T_{NUT}$ and a payload size l_i =46 bytes. The results of the simulations are shown in Fig. 6.6 for different network update times ranging from 4 ms...32 ms and a data rates of 12 Mbps.

Dynamic Retransmissions

The second scenario investigates the retransmission behaviour with respect to the packet loss rate experienced by the application. Different simulations have been

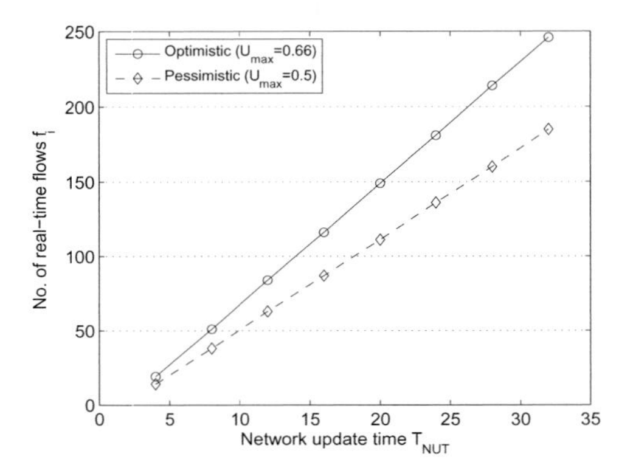

Fig. 6.6: Number of real-time traffic flows for different admission control reserves

conducted with an average packet error rate imposed by the wireless channel and varied from 0 . . . 20 %. A fixed number of 10 nodes and 20 real-time traffic flows are configured. The adaptive retransmissions are set to use a given maximum number of retransmission attempts N_{ret}. The simulations are run with $N_{ret} = 0, 2, 4$ and the resulting packet loss rate is analysed.

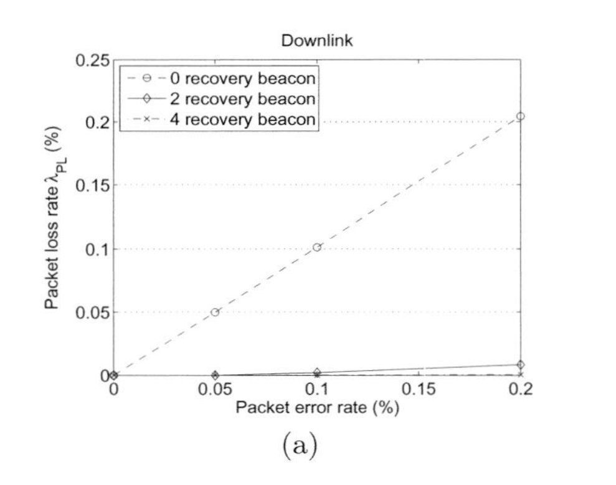

(a)

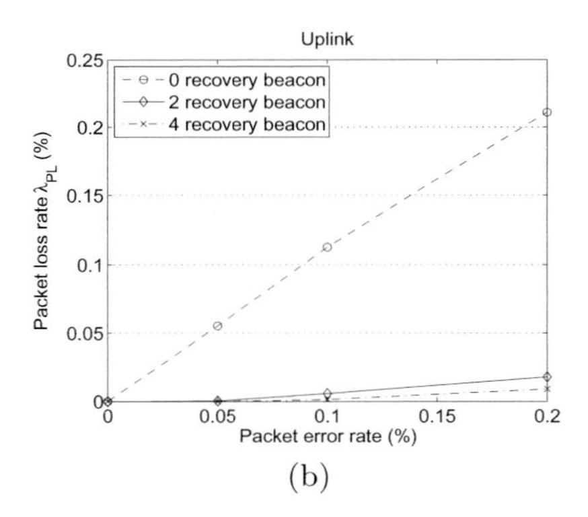

(b)

Fig. 6.7: Packet loss rate vs. packet error rate for a different number of retransmission attempts for (a) downlink, (b) uplink

The results are shown in Fig. 6.7 for the uplink and downlink direction. It is shown that the recovery mechanism is able to avoid packet losses almost completely up to a packet error rate of 10%. Even with a packet error rate increased to 15% and 20%, the packet loss rate is not exceeding 2%. Thus, the adaptive retransmission mechanism is able to increase the reliability of transmissions in the presence of frame losses.

7

Provision of a Global Time Base

In order to allow a synchronization of all IWN system components, including the wired communication network, a global time base must be established by means of clock synchronization. Establishing a system wide synchronization has two major advantages. First, different communication systems can be operated synchronously, i. e., the deterioration of the system capabilities, especially in terms of temporal characteristics, caused by asynchronously running communication systems can be reduced or even minimized. Second, real-time communication systems which rely on decentralized approaches, as the TDMA based medium access control proposed in Sect. 5, require the same notion of time within a given accuracy in all nodes of the system. For such a coordinated access, which is not only limited to TDMA approaches, a common notion of time between the wireless nodes is mandatory.

Providing a global time base pose two main challenges for the clock synchronisation within our proposed concept, (i) the clocks of the wireless system must be synchronised with a sufficient accuracy and precision, and (ii) the clock synchronisation of the wireless system must be integrated into the wired system, and interoperable to the wired clock synchronisation within the required accuracy. Both challenges are addressed in this chapter. First existing research activities and proposals are analysed for their suitability. Based on the obtained results, a wireless clock synchronization concept is proposed which seamlessly integrates into existing wired networks. Finally, the concept is analysed and evaluated based on a prototypical implementation.

7.1 Clock Synchronization in Hybrid Networks

Within this section the related work for clock synchronization in wireless networks is discussed and analysed with respect to its requirements.

7.1.1 Problems and Requirements

The most significant problems of clock synchronization are discussed in [40]. The behaviour of the network and the timestamping constribute to a clock reading error which influences the achievable accuracy. The clock reading error depends on the hardware and the used physical layer and can be summarized into the following factors which must be considered.

- Communication jitter in terms of varying delays of frame transmissions
- Clock reading error of outbound frames, referring to the timestamp taken when a frame is leaving the node. The timestamp should be taken as close as possible to the physical layer
- Clock reading error of inbound frames, this is identical to the previous point but for incoming frames
- Asymmetries caused by the communication channel, since the delay measurements rely on the assumption of having a symmetric characteristic

The different contributions of these error sources result in a given accuracy α and precision Π of a clock, according to [41] defined as follows.

Definition 7.1 (Accuracy) *A local clock C_p of node p runs with accuracy α_p, if $C_p(t)$ in an α_p-neighbourhood around the absolute time t during an observation period τ, meaning*

$$|C_{\mathrm{p}}(t) - t| \leq \alpha_p \qquad \forall t \in \{\tau\} \tag{7.1}$$

Definition 7.2 (Precision) *The precision Π of a clock is the upper bound of the difference between the the local clock C_p of node p and the local clock C_q of node q during a given observation period τ, such that*

$$|C_{\mathrm{p}}(t) - C_{\mathrm{q}}(t)| \leq \Pi \qquad \forall t \in \{\tau\} \tag{7.2}$$

As relevant metric for the accuracy α_p the mean clock offset is usually used. The precision Π is measured as the standard deviation of the clock offset [110]. In addition to this, the P5- and P95-percentiles are used in this chapter. On the MAC layer of the presented IWN a time slot resolution of 9 µs with an interframe space of 10 µs is used. Therefore, the accuracy of the distributed local clocks must be well below, i. e., in the range of $\leq$ 2 µs, to avoid any negative influence on the TDMA-based medium access. If this requirement is not met by the clock synchronization, the distributed IsoMAC node will start their transmissions too early or too late resulting in collisions and a violation of the real-time guarantees.

7.1.2 Related Work

In the case of 802.11, the internal synchronization is based on the Timing Synchronization Function (TSF). Every WLAN infrastructure network employs the TSF in order to keep all nodes in the network synchronized. Every node maintains its local time in a 64 bit TSF timer with a resolution of approx. 1 µs. The value of the timer at the node is corrected according to the time of the AP which is sent by the AP to the clients inside beacons on a periodic basis. The nodes receive the beacon, read the contained TSF timer value of the AP and replace their respective timer values with the value from the beacon. However, TSF is used for internal WLAN management and has no relation with the overall system clock. Therefore, this functionality cannot be used for system-level applications, because it provides only relative time information in the sense of a timer rather than a clock and no global time base can be established on the node side. Moreover, TSF is implemented in hardware without being accessible in desirable way. In the future, the 802.11v amendment will provide timing measurement frames, which might allow to establish a global time base, but the required HW features are not available [109]. Hence, TSF based synchronization will not be of any use for this work.

This prompts the need for an application layer protocol to provide highly accurate synchronization for WLAN entities. The IEEE 1588 PTP [68] is a promising candidate, because it has been widely used for synchronization over Ethernet. Nevertheless, in many scenarios COTS WLAN devices shall be used to minimise the cost of the communication setup. In this case a modification of the nodes' hardware is not possible due to closed and fully integrated chipsets. Hence, software timestamping inside the protocol stack remains the only option for clock synchronization. In order to deal with these issues, the combination of an application layer synchronization protocol and software timestamping over WLAN has been a topic of interest in several existing research works.

An IEEE 1588 implementation over WLAN using software drivers has been conducted by Kannisto et al. [88] with a synchronization accuracy of 4.6 µs and variance of 2.56 μs^2 for infrastructure mode. However, these results have been obtained from averaging 10 test runs, each of which lasted only 10 minutes with a synchronization interval of 2 s. This small sample space is not enough for establishing valid results about the accuracy of the proposal and will not be further considered in this work.

Cooklev et al. [25] considered the IEEE 1588 [68] standard within wireless networks. In general, IEEE 1588 is intended for industrial usage and high-accuracy implementations. They investigated the upper bound of the achievable accuracy when using IEEE 1588 for synchronization over IEEE 802.11b [70]. The paper investigates potential accuracy limitations introduced by the physical layer of the IEEE 802.11b wireless local area network. Some experimental results show that the contribution of the physical layer to the degradation of the clock synchronization accuracy is several hundred nanoseconds. These results can be considered as the lower bounds of the possible precision when using IEEE 1588. This is due to fact, that time-stamping can not be done before the interface of the baseband processing and the MAC controller. This study does not analyse an actual implementation. Instead, it spotlights potential instants where hardware timestamping can be done to provide synchronization accuracy in the range of nanoseconds and provides many other useful insights.

Bang et al. [3] have provided a PTP implementation over IEEE 802.11n for multimedia services using software support and statistical filtering. However, the final synchronization jitter is 400 µs and not relevant for this work. Another approach for industrial and home networks is presented by Cena et al. in [18] and based on a reference broadcast infrastructure synchronization protocol. The approach seems to be promising, but the achieved accuracy is above 10 µs.

Clock synchronization based on WLAN device drivers has been investigated for infrastructure mode by Mock et al. [119, 120]. A final synchronization precision of 150 µs has been achieved using a proprietary driver in a Windows environment. The authors propose a mechanism to broadcast time information for a higher precision by reducing the error sources during transmission. By shortening the time-critical path for a synchronization packet, a decoupling of the timestamping from the actual frame transmission is achieved, leading to an increased precision. Due to constraints of the available hardware at the time of this work, better results could not be achieved. However, the presented concept is rather interesting, because it has several similarities to the *Follow_ Up* principle which is deployed in this work. Furthermore, the fault-tolerance of the proposal is an interesting property of the protocol.

The authors of [108] propose an approach to optimize synchronization accuracy for IEEE 1588 over WLAN by modifying the PI controller of the clock servo. Two main error sources are identified which is the timestamping error and the oscillator noise. Both error sources are analysed and used as input for a simulation model of the clock servo. Using the model the constants K_p and K_i of the PI controller for an optimal synchronization accuracy are derived. An optimal synchronization performance is achieved if the software timestamps are considered less important by the clock servo. Instead it relies more on its oscillator. The results of this work promise a very high accuracy.

Discussion After the analysis of related work in the area of clock synchronization, it is concluded that to the best of the authors knowledge none of the existing approaches so far is able to provide the required clock synchronization accuracy. However, the work of Mock et al. [119] as well as the work of Mahmood et al. [108] will be further considered and integrated into the proposed clock synchronization concept of this work.

7.1.3 Precision Time Protocol

In the context of industrial automation the IEEE 1588 Precision Time Protocol (PTP) is one the most widely distributed protocols [17]. This is due to its achievable high precision and the relatively simple protocol. PTP mainly relies on a very accurate hardware-based timestamping of received and transmitted frames. The timestamp represents the exact point in time either when a *Sync* packet left the device or it is received. This leads to an achievable synchronization precision of a few nanoseconds.

The protocol will be further explained using Fig. 7.1 which reflects the PTP principle. The master sends a *Sync* packet at time reaching the slave at t_{S2}. Using hardware timestamping the slave can calculate the difference $t_{M1} - t_{S2}$ and synchronize to the master. If hardware timestamping is not supported the master sends a *Follow_ Up* packet after the *Sync* packet which contains the timestamp t_{M1}. The offset correction is done at the slave after reception of the *Follow_ Up*.

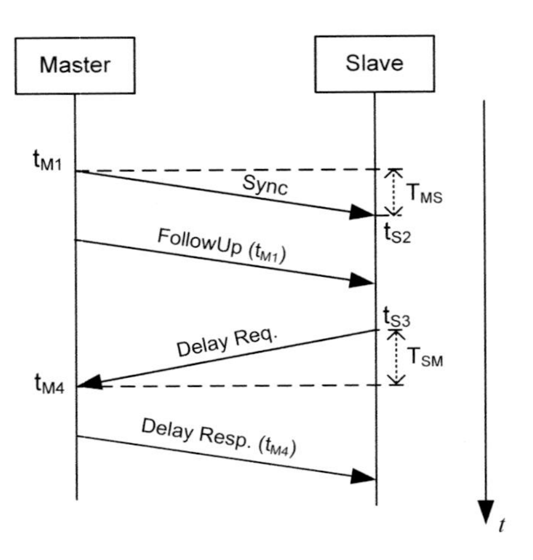

Fig. 7.1: Wireless clock synchronization concept

Since this calculation includes the transmission time from master to slave T_{MS}, a constant offset of both clocks would be the result.

This issue is solved by sending a *Delay_Req* packet at time t_{S3} from the slave which is received at t_{M4} after the transmission delay T_{SM} at the master. The master replies with a *Delay_Req* packet and enables the slave to calculate the transmission delay T_{Δ} with Eq. (7.3). However, this requires that the transmission delay in both directions is symmetric, i. e., $T_{MS} = T_{SM}$.

$$T_{\Delta} = \frac{(t_{S2} - t_{M1}) - (t_{M4} - t_{S3})}{2} \tag{7.3}$$

In PTP different types of clocks are specified, four of them being relevant for this work. The *grandmaster clock* is responsible for providing the global reference time, it is usually an ordinary clock with the most precise clock in the network. *Transparent clocks* reside in nodes with more than one port and compensate their internal delay when forwarding frames. *Boundary clocks* are also characterized by having two or more ports, but they act as slave in one network and are the clock master of another network. Finally, *ordinary clocks* reside in nodes with a single network interface and can be both master or slave.

7.2 Wireless Synchronization Concept

In hybrid networks, consisting of wired and wireless systems as introduced in Sect. 4.2, the final synchronization accuracy is affected by the performance of both parts of the system. For the synchronization of the wired part it is very common to use IEEE 1588v2 with hardware timestamping [17] or derivatives of it, such as PTCP. The achievable synchronization accuracy for these systems is in the range of a few nanoseconds [105].

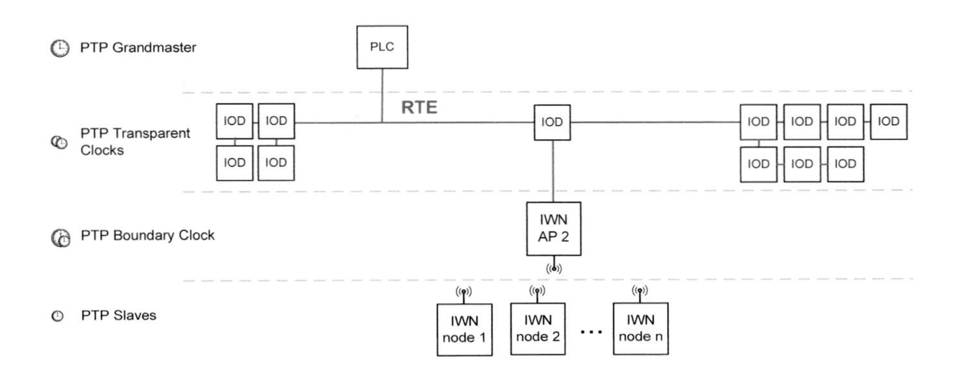

Fig. 7.2: Clock synchronization in hybrid networks

Hence, the clock synchronization accuracy for the hybrid system is mainly determined by the wireless components, because only software timestamping is supported. The hybrid network and its clock hierarchy is shown in Fig. 7.2. The PLC will act as grandmaster clock for the hybrid network. The wired IODs are usually transparent clocks, i. e., they calculate their bridge delay and add this time to the *Sync* packet as a separate field before forwarding it. Following the integration proposed in Sect. 4.3, the APs act as boundary clocks. The AP will act as a slave to the grandmaster, but it will be the master of its wireless network. Having such cascaded clocks increases the end to end jitter and degrades the synchronization accuracy between grandmaster and wireless nodes. However, since the wired jitter is in the nanosecond range, it is very small compared to the wireless synchronization and the influence of cascaded clocks can be neglected. Therefore, the wired synchronization will be completely based on IEEE 1588v2 without any modification and the wireless part is modified accordingly.

According to the previous analysis a wireless synchronization concept based on PTP will be presented in this section and shown in Fig. 7.3. The concept is based on the ^{flex}WARE concept for wireless clock synchronization [111, 112], but has been extended with several modifications to address the specific needs of the IWN solution approach. Furthermore, an improved clock servo is used from the work of Mahmood et al. [108]. The clock servo explicitly models the error sources described in Sect. 7.3, which results in adapted servo parameters and an optimized synchronization accuracy.

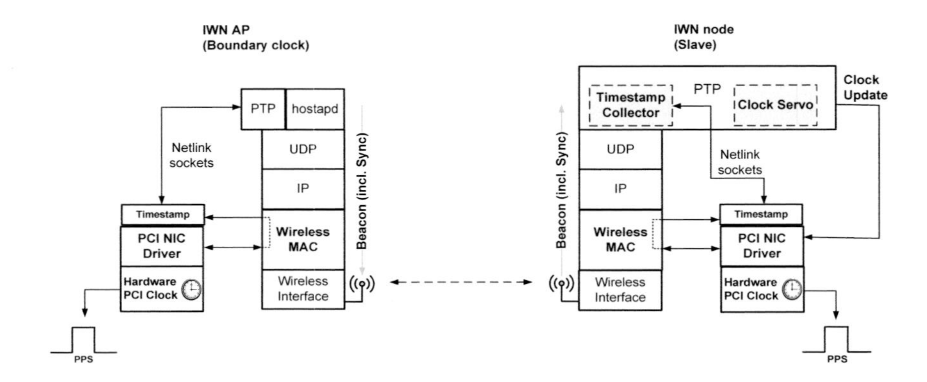

Fig. 7.3: Wireless clock synchronization concept

Clock synchronization start-up The whole start-up phase of the wireless clock synchronization system is shown in Fig. 7.4. The wireless nodes are always initialized through the standard authentication/association process and through implicit prior approval by the admission control procedure as described in the previous section.

After the node is successfully authenticated and associated to the AP, an IP address is assigned. As soon as the node has its own IP address, it starts to receive synchronization messages from the AP master clock. It is assumed that the AP is already synchronized at this time, because it only accepts nodes to associate with it when the AP is already synchronized.

Once the node is properly synchronized to the AP master clock, the new synchronization status is sent to the AP. After this the IsoMAC operational phase starts, i. e., the real-time traffic flows between the AP and the node are exchanged. The whole wireless start-up procedure is repeated for new nodes entering the system.

7.2.1 Precision Time Protocol over Wireless

In the simplest case, PTP messages can be directly sent via WLAN and the resulting synchronization accuracy can be monitored. This scheme is very similar to the one introduced in [155] where a normal wired PTP stack has been used, but the wired medium is replaced by a WLAN link. However, this scheme led to a poor accuracy, because of the random medium access delays in WLAN which will compromise the accuracy of the timestamps.

To avoid this, software timestamping can be done inside the interrupt handling routine of the driver. A hardware interrupt will be raised, whenever a packet is successfully transmitted or received. After this the host computer will be prompted to draw a timestamp. The timestamps will be routed later on to the PTP stack and can be used for offset and delay compensation. In addition to providing these

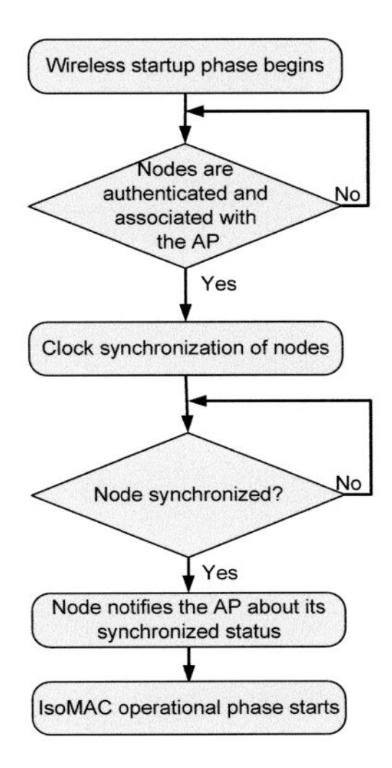

Fig. 7.4: Start-up phase of the wireless clock synchronization

timestamps from the kernel space to the PTP stack, the following two general issues arise when using PTP over WLAN.

7.2.2 Clock Synchronization Using Beacon Frames

A further step to increase the efficiency of clock synchronization in WLANs is the usage of beacon frames to transmit application layer synchronization information. This leads to a reduced bandwidth consumption as compared to the inevitably arising communication overhead for clock synchronization. A similar concept has been proposed in [119] and is extended in this work for using PTP for clock synchronization.

In the proposed IWN the AP sends periodically beacon frames as layer 2 broadcasts to reveal its capabilities, and to send the dynamic communication schedule to the nodes. Beacon frames can also contain vendor specific information as tagged data [70]. Hence, they are a suitable alternative to ordinary *Sync* messages as specified in the PTP standard.

In order to add a PTP compliant timestamp to the beacon frame, an additional amount of 22 bytes is needed in the vendor specific field. Despite the lower transmission rate of beacon frames (6 Mbps), the additional bytes consume less time than actually sending out separate frames that include all protocol overhead of the PHY and MAC layer such as preamble, interframe spaces, and header information. Furthermore, the fault-tolerance approach of [119] is also integrated into this concept. Instead of embedding only the most recent beacon transmission timestamp $t_{B_n\ 1}$ into B_n, the j previous timestamps up to $t_{B_n\ j}$ are also embedded. This allows to tolerate omission failures up to a maximum omission degree OD_{max} as assumed in our system model, if $j \geq OD_{max}$.

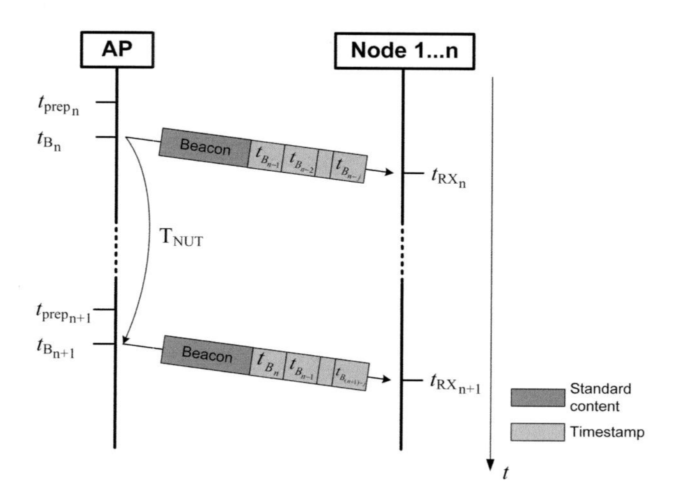

Fig. 7.5: Principle of timestamp transmission using beacon frames

The timestamps will be taken by the device driver module of the wireless MAC as described in Sect. 7.2. The timestamp propagation principle from the AP to all associated nodes is shown in Fig. 7.5. The time instants shown in this diagram indicate all relevant actions performed in the driver module, that involve timestamping and forwarding. At time t_{prep_n} the standard content of the next beacon frame is prepared, and afterwards extended by the beacon transmission timestamp $t_{B_{n\ 1}}$ of the previous beacon, which was taken by the drivers interrupt handler routine.

This is similar to the *Follow_up* principle envisaged by the PTP standard for transmitter hardware not capable of on-the-fly timestamp insertion. Certainly, an additional delay $T_{NUT} \leq 100\,\text{ms}$ is introduced for the offset calculation, because $t_{RX_{n+1}}$ is the earliest point in time when the clock offset of t_{RX_n} to t_{B_n} can be calculated. This is a contradiction to the *"as soon as possible"* principle stated in the PTP Standard for handling *Follow_up* messages. However, timestamping within the beacon interval of $< 100\,\text{ms}$ seems to be short if a clock drift of $\leq 1\,\text{ppm}$ is considered. The proposed mechanism is further analysed by analytical calculations and empirical measurements in Sect. 7.4 with respect to accuracy and precision.

7.2.3 Path-delay Asymmetry Compensation

IEEE 1588 assumes in its path-delay calculation that the master-to-slave delay and the slave-to-master delay are symmetric. However, this assumption does not hold true and can lead to a huge bias in the final accuracy. In wired communication, this asymmetry comes from the physical layer of the devices and from the cable itself. For wireless communication using interrupt-based software timestamping, a major source of this asymmetry is multicasting of packets from a node to the AP in infrastructure mode. Inside PTP, for the downlink direction (from the master to the slave), a *Sync* packet is sent as a multicast packet to the nodes. This *Sync* packet will be transmitted via the wireless channel and received by all nodes in the wireless network. The AP and the nodes will draw the timestamps upon successful transmission and reception of the packet respectively. As *Sync* packets are sent as multicasts, a corresponding acknowledgment (*ACK*) will not be sent from the receiver as defined in [70].

Since a direct communication between nodes is not allowed in an infrastructure WLAN, *Delay_ Req* packets cannot be transmitted as multicast directly to all nodes in the network in the uplink direction (slave to master). The AP has to receive the multicast packet from the node and has to forward it to all nodes in the cell. This also includes sending the packet back to the node who has initially sent the *Delay_ Req* packet in the first place. It is obvious that this forwarding of *Delay_ Req* packets by the AP to all nodes only wastes bandwidth, as the nodes will discard these packets.

Hence, the *Delay_ Req* multicasts are replaced by unicast messages resulting in another problem. Upon receiving the packet, the receiver must send an *ACK* in response before drawing the receive-timestamp. As no *ACK* has been present in the downlink direction, the slave-to-master path-delay is now greater than the master-to-slave path-delay. The constant size of *ACK* frames is 112 bits resulting in a path-delay asymmetry of 44 µs, assuming that the *ACK*s are sent at the basic rate of 6 Mbps. Moreover, the length of a short interframe space [70] has to be added. This asymmetry will be reflected in the final path-delay calculation and therefore the clock synchronization accuracy will have a constant offset. In order to remove this asymmetry, the *NoACK* policy should be used for *Delay_ Req* packets. An even more efficient usage of available bandwidth can be achieved by using Beacon frames for synchronization as described in the next section.

7.3 Challenges for Time Stamping

As mentioned in the previous section and as discussed in [111], software timestamping inside the WLAN driver will be employed in this work. In the interrupt based synchronization scheme, timestamping delay and jitter are mainly confined to the sources mentioned below.

7.3.1 Packet Reception Time

This is the time taken by the receiver to receive the packet once the transmission has actually begun. This delay pertains to the wireless propagation of the packet and the functioning inside the physical layer of WLAN. Besides factors like settling of the gain controller, timing and phase synchronization, and stabilization of the equalizers, it also depends on random wireless channel fluctuations. The quantification of this parameter is difficult as the wireless channel is never stable. For instance, the jitter of the packet reception time in a stable line of sight (LOS) connection is found to be 120 ns [111]. Though facing poorer radio conditions, over longer distances, and in the presence of interference, this jitter is bound to fluctuate. If the radio conditions severely change in a short span of time, it might happen that the path-delay from master-to-slave differs from the slave-to-master delay. This can lead to channel asymmetry which has to be tackled by other means, e.g., statistical filtering or sending *Delay_ Req* packets at the same rate as *Sync* messages. Nevertheless, disturbances of the wireless channel can affect timestamping precision and their effect needs to be quantified and compared with other sources of jitter present in synchronization.

7.3.2 Interrupt Notification

The interrupt notification time is the time elapsed between the complete transmission or reception of a WLAN packet and the time to notify the host CPU via an interrupt that a packet has been successfully received or transmitted. It is impossible to determine this time precisely as this would require a direct access to the WLAN chipset hardware which is not provided for COTS chipsets. In [109] the delays of the transmitter path and the receiver path are characterised.

The interrupt notification time of a packet at higher data rate has been smaller than at lower data rates albeit with the same packet size. This shows that the interrupt notification obtained from this experiment includes some of the packet reception time as well. This also reinforces the concept that longer packets will stay longer inside the WLAN chipset as compared to packets with smaller payload. Hence, if *Sync* and *Delay_ Req* packets are of different sizes, this will lead to an asymmetry in path-delay calculation which will depend upon the size of the individual packet and the data rates at which they are sent. Though in PTP, *Sync* and *Delay_ Req* packets are of similar length, if beacons are to be used as *Sync*, then it has to be ensured that *Delay_ Reqs* are of similar size as that of beacon and are also sent at the same data rate as of a beacon.

Nevertheless, the jitter of the interrupt notification time is an important parameter which can influence the final synchronization accuracy. The investigations of [109] show that this interrupt notification time has a mean value of 1.31 µs with a standard deviation of 46 ns for the transmitter side, and a mean value of 8.9 µs with a standard deviation of 110 ns for the receiver side.

7.3.3 Interrupt Handling Delay

This delay is the time taken by the CPU to handle the hardware interrupt and to start the interrupt handling routine for the appropriate hardware interrupt. This delay depends on many factors such as number of CPU cores, CPU speed, CPU frequency scaling etc. Even though no measurements have been taken for this delay, in [144], it has been shown that jitter in interrupt handling delay is

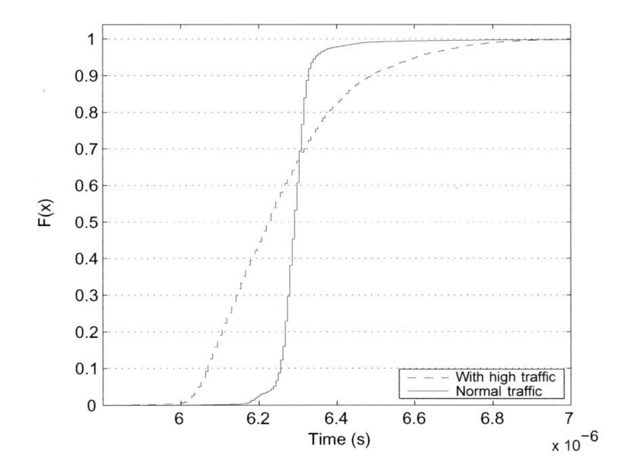

Fig. 7.6: ECDF plot of the elapsed time to draw a timestamp

300 ns. This delay increases with CPU load and the maximum interrupt delay can be as high 70 µs. Nevertheless, such big delay values can be filtered out in the final synchronization algorithm to avoid such huge deviations during clock adjustments.

7.3.4 Timestamping Delay

The timestamping delay corresponds to the time it has actually taken to draw the timestamp. To measure this delay, a measurement setup similar to [10] is established using an industrial PC with a 2.4 GHz CPU, similar to the one used in the last section. The timestamping is done by reading a UTC time register residing on a customised PCI card running at 62 MHz. The time taken by the system to read this register is measured by the *rdtsc()* interface which is very fast and has a very low jitter. To measure the jitter for timestamping, the timing register is read under normal conditions and under high WLAN traffic for at least three hours each. The case for reading the register under high CPU load is skipped, because of optimization of *rdtsc()* interface as mentioned in [10]. Figure 7.6 shows the empirical cumulative distribution function (ECDF) plots of timestamping delay under nominal conditions and under high network load. The median in case of normal conditions is 6.23 µs while under network load is 6.29 µs. However, the standard deviation of timestamping delay is 60 ns while in case of a loaded network it increases to 175 ns. This shows that even in a loaded network, the jitter remains in the low nanosecond range and does not significantly affect the final synchronization. It should be noted that some filtering has been carried out on these measurements to remove obvious outliers. These outliers have been more frequent in the loaded network than in the unloaded one. They further highlight the need for filtering before using the timestamps to control the clock.

7.4 Accuracy and Precision Analysis

In order to evaluate the achieved improvements of the proposed concept, an analytical throughput estimation is done and one experiment consisting of three different scenarios for the accuracy and precision analysis is conducted. After introducing the testbed setup, the increased efficiency of the proposal in terms of throughput is investigated and compared with the system using ordinary *Sync* frames, followed by the accuracy and precision analysis.

7.4.1 Testbed Setup

The testbed setup consists of one IWN AP and one IWN node establishing a connection. The AP can be considered as boundary clock and is the default master clock for the wireless segment. The node acts as default slave clock. The hardware consists of two industrial PCs with a 2.4 GHz CPU and a Linux operating system.

Implementation

The *ath5k* Linux device driver has been utilized for wireless communication which supports chipsets from Atheros and acts as an interface between *mac80211* and the WLAN chipset. In general, two types of wireless cards exist, (i) cards with a fullMAC chipset and no possibility to modify the MAC layer protocol, and (ii) cards relying on a softMAC chipset. A softMAC chipset manages the MAC Layer Management Entity (MLME) completely in software. The Atheros *AR9XXX* and *AR5XXX* families are examples for this type of chipset [102]. The MLME is currently implemented in user space, where different user space modules realize the needed functionality for the AP and the node. The *cfg80211* is the new Linux wireless configuration API, while the *nl80211* is used to configure a device. The *nl80211* is a new 802.11 Netlink interface public header, which is used to transfer information between kernel modules and user space processes. It consists of a standard socket based interface for user processes and an internal kernel API for kernel modules (*mac80211*). The *mac80211* is the Linux API used to write softMAC wireless drivers. *mac80211* is a wireless subsystem which resides in kernel space and provides interfaces to different driver modules depending on the specific wireless interface hardware. The *nl80211* features are enabled in *cfg80211* to provide a communication transport between both domains. The *mac80211* stack is also responsible for communicating with the WLAN driver which communicates directly with the hardware.

Scenario

In order to remove the path-delay asymmetry, several changes have been made inside the *ath5k* driver and *mac80211* stack. The actual clocks to be synchronized are adder-based clocks which reside on the network interface cards and can be accessed by PTP from the application layer. The first scenario has the objective to show the influence of the path delay compensation. Hence, in this scenario, the *Delay_ Req/Delay_ Rsp* packets are not sent and no compensation for the path delay is possible. The *Sync* information is embedded into the beacon frame and sent with T_{Sync} =64 ms. In the second scenario, the beacon based mechanism with the same sync rate is used, but the path delay compensation is now enabled. In the third scenario, the beacon interval, i. e., T_{Sync}, is modified from 64 ms to 1000 ms. The path delay compensation is again active, and *Delay_ Req/Delay_ Rsp* packets are sent.

The obtained synchronization accuracy is recorded by comparing the 1 pulse per second (PPS) output from the master and the slave clocks on an oscilloscope. The mean clock offset and its standard deviation are determined. The used metrics have been selected, because they are frequently used to characterise the accuracy and the precision of synchronized clocks.

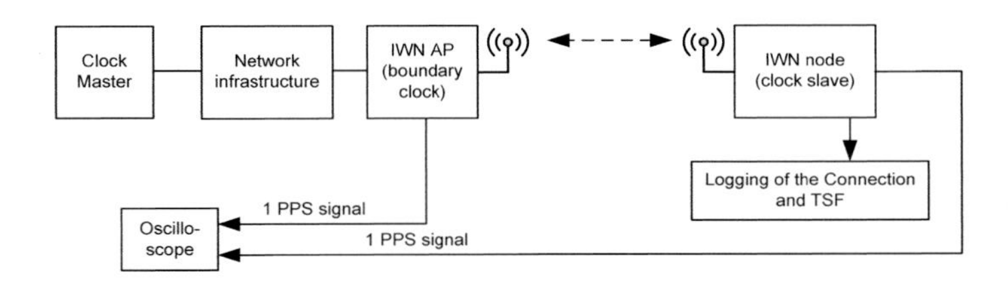

Fig. 7.7: Testbed setup for the clock synchronization evaluation

7.4.2 Bandwidth Utilization

One advantage of using beacon frames to transmit synchronization information is an increased efficiency of clock synchronization in WLANs. This leads to a reduced bandwidth consumption avoiding inevitably arising communication overhead for clock synchronization. In this subsection, the normal approach of sending *Sync* frames is compared against the Beacon based approach in terms of throughput B.

The resulting bandwidth utilization B for sending separate *Sync* frames is given in Eq. (7.4).

$$B = \frac{l_{Sync} + l_{FollowUp}}{T_{Sync}} + \frac{l_{Dly}}{16} \tag{7.4}$$

where l_{Sync} is the amount of data in bit of a *Sync* packet including all protocol overhead, $l_{FollowUp}$ is the amount of data in bit of a *Follow Up* packet including all protocol overhead, and l_{Dly} the amount of data for the delay request/response mechanism.

The additional throughput for the Beacon based synchronisation is given in Eq. (7.5)

$$B = \frac{l_{BeaconMod}}{T_{NUT}} + \frac{l_{Dly}}{16} \tag{7.5}$$

where $l_{BeaconMod}$ is the amount of additional data in bit for embedding the *Sync* information into the beacon frame, T_{NUT} the configured network update time, and l_{Dly} the amount of data for the delay request/response mechanism.

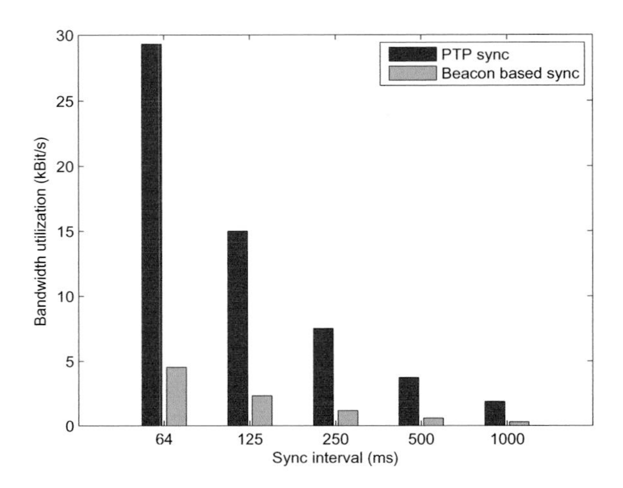

Fig. 7.8: Resulting bandwidth utilization of separate *Sync* frames compared to the proposed beacon-based approach

The increased efficiency of the beacon-based approach, as compared to sending separate *Sync* frames, is shown in Fig. 7.8. The sync intervals T_{Sync} are varied from 64 ms to 1000 ms. Sending separate *Sync* frames results in an additional bandwidth utilization which is 6 times larger than for the beacon based mechanism, considering the largest sync interval of 64 ms. Depending on the deployment scenario of the solution, a considerable amount of bandwidth is conserved when using the beacon based approach. Hence, it is available for transmitting real-time critical traffic flows.

7.4.3 Accuracy and Precision

The obtained results have been measured during the stable phase of PTP, i. e., results from the initial settling period of PTP have been ignored. A confidence level of 95% has been chosen, resulting in sample size of $N = 30000$. Some obvious outliers in the 1 PPS comparison, resulting from non-real-time behaviour of the Linux operating system, have also been removed by *a posteriori* filtering.

The results of the first and second scenario are shown in Fig. 7.9(b) and Fig. 7.9(a) respectively.

The histogram in Fig. 7.9(a) shows the synchronization accuracy achieved only with the help *Sync* information. The mean offset in synchronization is found out

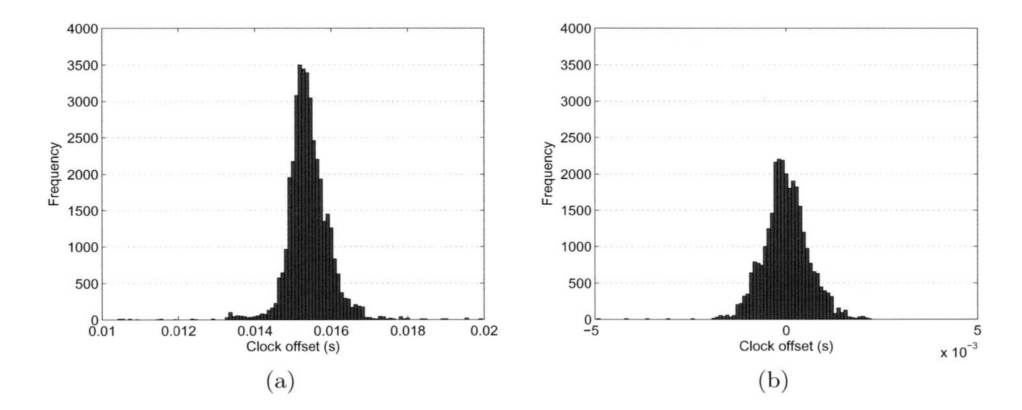

Fig. 7.9: Clock offset with and without delay asymmetries caused by the acknowledgment

to be 15.402 µs with a standard deviation of 0.565 µs, and 81% of the values are in the 1σ-interval $[14.837, 15.967]$ µs. Fig. 7.9(b) shows the synchronization accuracy when utilizing *Delay_ Req* messages. In this case, the synchronization error in terms of mean offset is found to be -0.257 µs with a standard deviation of 0.680 µs. In the second scenario 79% of the values are within the 1σ-interval $[0.937, 0.423]$ µs.

Even though the *Delay_ Reqs* mechanism removes the bias in synchronization, the obtained synchronization jitter is slightly higher. This is due to the usage of two more software timestamps for the *Delay_ Reqs*/*Delay_ Resp* packets, their presence adds further noise to the jitter. If it desired to minimize this error, it is recommended use only *Sync* messages along with statistical means and fully avoid using *Delay_ Req* packets as the changes in the path-delay, mainly caused by the wireless channel, are minimal and the delay sources reside mainly in the host PC.

The third scenario is carried out to analyse the impact of an increased sync rate on the precision of the wireless synchronization using the beacon based clock synchronization. As in the previous measurements, some outliers are removed from the sample space, i. e., only samples within the interval $[-5, 5]$ µs are for used for the analysis.

The resulting clock offset is shown in Fig. 7.10 as CDF for different sync intervals T_{Sync}. The CDF has been derived by a normal distribution fitting using the well-known *Freedman-Diaconis rule* [38]. A summary of the results with its most important statistical parameters is provided in Table 7.1. The sync interval is given in column one. The second and the third columns provide the mean clock offset and the standard deviation of it, representing the metrics for the synchronization accuracy. The last column lists the percentage of outliers being exempted from the sample space. The third and the fourth column provide the P5- and P95-percentile of the measurements respectively.

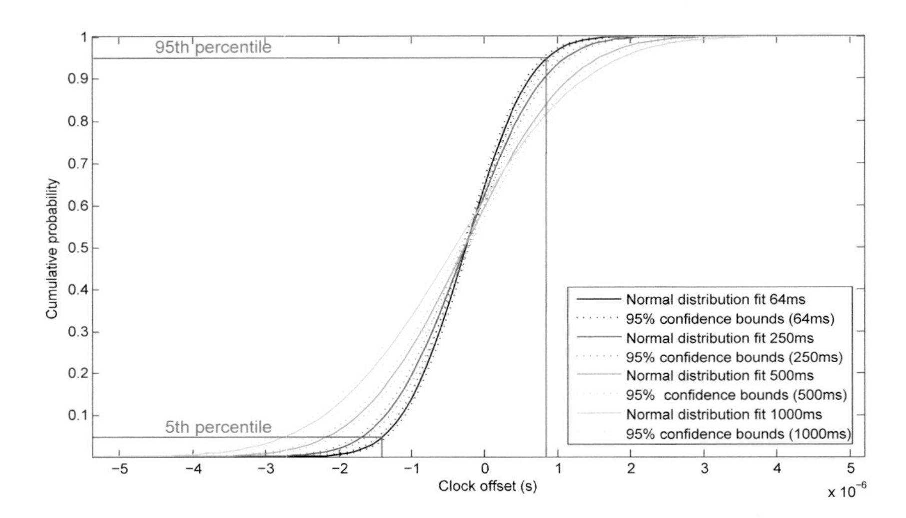

Fig. 7.10: CDF of the clock offset for different sync intervals T_{Sync}

Table 7.1: Clock offset for different *Sync* intervals

Sync interval T_{Sync} (ms)	Mean (µs)	Std. dev. (µs)	P5-perc. (µs)	P95-perc. (µs)	Outlier (%)
64	-0.2574	0.6809	- 1.48	0.85	0.20
250	-0.2732	0.8432	- 1.76	1.1	0.25
500	-0.2769	1.1277	- 2.15	1.62	0.41
1000	-0.4091	1.3933	- 2.85	1.9	0.42

For the sake of simplicity of a comparison among different sync intervals, the percentiles have been calculated to highlight the relationship between synchronization accuracy and sync interval. Both values are also shown for T_{Sync}=64 ms in Fig. 7.10. The results show that the standard deviation is minimal for T_{Sync}=64 ms with a value of 680 ns and 90% of the values are in the interval $[-1.48, 0.85]$ µs. In this measurement 0.2% of outliers are filtered out. Whereas for the standard mechanism and a sync interval of T_{Sync}=1000 ms, the standard deviation is doubled to 1.393 µs and the 90% interval also increases significantly to $[-2.85, 1.9]$ µs.

As a result of this analysis, it can be stated, that embedding the *Sync* information into the ongoing Beacon transmission is highly recommended. Mainly because of two reasons:

- The resulting protocol overhead for clock synchronisation is limited to the transmission of *Delay_Req* and *Delay_Resp* frames. Their transmission rate is usually 4 to 16 times lower than the transmission of *Sync* frames. This is especially important in the case of wireless networks, where the available bandwidth is limited.

- A higher precision can be achieved, if the sync rate is increased. This is automatically achieved by using the beacon frame and its corresponding interval T_{NUT}, and no additional trade-offs are associated to this. Comparing the 5th- and the 95th-percentile, the results are twice as much with the usual T_{Sync} =1000 ms as compared to the beacon-based approach with $T_{Sync} = 64$ ms.

7.5 Concluding Remarks

The presented concept for wireless clock synchronization is based on the ordinary PTP protocol stack and software timestamping using a modified open source driver for a COTS WLAN chipset. It is shown that this approach is able to achieve accuracies below 2 µs in terms of mean clock offset and its associated standard deviation, despite several error sources introduced by the software timestamping mechanism.

Moreover, the sync rate should be kept as high as possible in order to achieve a higher precision. Using the standard PTP, separate *Sync* packets with a highly increased sync rate would utilize a considerable amount of wireless bandwidth for synchronization traffic, and could result in an additional delay experienced by the *Sync* packets which further decreases the synchronization performance. Therefore, beacon messages are perfectly suited to increase the sync rate for a higher precision of the synchronization. Even though they slightly decrease the accuracy, the usage of *Delay_Req* packets is necessary for path delay compensation. Statistical means would be an alternative which could be applied to achieve a path-delay compensation without using *Delay_Req* messages, but would probably require a calibration of the whole system before deployment.

The main conclusions for the provision of a global time are that the achievable accuracy of the presented PTP based approach is sufficient for the TDMA MAC proposed in Ch. 5, as well as for the application. Further, the proposed modification of embedding sync information into beacon frames instead of sending separate sync frames allows a higher sync rate for improving the precision of the clock synchronization. The mechanism is also superior in terms of bandwidth utilization.

8

Evaluation

In this chapter the key components of the presented solution approach of an Isochronous Wireless Network (IWN) are evaluated to prove its validity based on our main application scenario. The experiments are conducted based on a prototypical implementation and a simulation case study.

After presenting an overview of the evaluation and highlighting their specific structure, the motivation for using both experiments and simulations is provided in Sect. 8.1. In the next section, the prototypical implementation is described. A description of the simulation model follows in Sect. 8.3. This section also includes the description of a new wireless channel model. The model is based on the channel characterisation in Sect. 3.2, and specifically developed to obtain more realistic simulation results, since the correct modeling of the lower layers is very important when simulating wireless networks. The simulation and the new channel model are validated using the prototypical implementation. In Sect. 8.4 the first case study of a reconfigurable manufacturing system is carried out in a real manufacturing environment using the prototypical implementation. In Sect. 8.5 the evaluation results of the second case study are presented based on the simulation model.

8.1 Overview and Structure

In contrast to other evaluation approaches, the presented solution is evaluated using two distinct methods, discrete event simulations and a prototypical implementation of the system. Both methods are contributing with different findings. Simulation based experiments allow to have an arbitrary number of nodes and the ability to collect all data you need from every node at any time. Whereas experiments with a real system are always limited in size. However, when evaluating a real system several advantages arise as well. The simulation models are always an abstract model of the real system. It might be not possible to model the entire real system as required. The main restrictions and advantages of both methods are listed in Table 8.1.

The first case study considers the application category of reconfigurable manufacturing systems. The experiments are carried out in a real manufacturing en-

Table 8.1: Comparison of simulation and prototype

Characteristic	Simulation	Prototype
No. of nodes	no limit	up to 10
Physical layer	model	real
MAC	accurate model	limitations due to internal timing
Wireless channel	abstract model	real
Communication stack	modeled latency	real latency
Synchronization	modeled error	real error
Results collection	log	expensive setups

vironment which is used for teaching purposes. This category is characterised by its challenging wireless channel conditions due to a dynamic open environment. Besides the varying channel, which is caused by moving machines or personnel, the environment and the wireless system is considered to be open for other wireless nodes, i. e., the system has to deal with a number of uncontrollable best-effort nodes, e. g., for monitoring and maintenance, which should not disturb the real-time communication. The temporal requirements of this application category are of real-time class 1.

The application category for the second case study is the wind energy automation considering the closed-loop control of the pitch motors and their video surveillance system. The wireless channel of this application category is rather static and only influenced by the rotation of the system. The application requires only a few real-time nodes, but the temporal requirements refer to real-time class 2. In comparison to the first environment, no other best-effort nodes are present other than those for transmitting the video stream for surveillance.

The evaluation of the presented IWN is structured as follows. In the first step, a brief introduction of every experiment, its objectives, and the relevant metrics are provided. In the second step, the evaluation approach and the implementation of it are described. In the third step, the obtained results are presented and discussed. Finally, the results of the evaluation are concluded and the main findings are highlighted. In general, both case studies consist of two experiments and follow the same evaluation approach:

- Experiment 1: Validate that the application requirements can be fulfilled by the IWN
- Experiment 2: Show the necessity of the components, which are required for the application category

In the first experiment, it is shown that the presented IWN is able to fulfil the requirements of the application. In the second experiment, the necessity of different components of the IWN is proven, by comparing results with activated and deactivated components. The selected components are based on the requirements of the application category. Thus they are different for both case studies.

8.2 Prototypical Implementation

For the first case study, an implementation prototype of the IWN solution approach is realised based on an industrial PC platform and a COTS WLAN hardware. Its internal architecture, including all relevant components for the evaluation, is briefly introduced in this section, separated into hardware- and software-specific parts. The system is shown in Fig. 8.1, more details are provided in [167].

Fig. 8.1: Industrial PC platform and WLAN hardware for the implementation prototype

8.2.1 Implementation Architecture

The whole implementation prototype is able to provide isochronous real-time communication between distributed automation devices using the IWN wireless infrastructure. The AP acts as a bridge between the wired and the wireless medium and several nodes might be connected to every AP. The nodes interconnect automation devices through a wireless link with the AP and the backbone network.

Even though both devices, FAP and FN, comprise the IWN components, their implementations are slightly different. The AP is responsible for controlling the medium, and must have certain functionalities such as the admission control, schedule distribution, retransmission handling, etc. The nodes only need a reduced functionality. They must be able to perform the admission, receive and process the schedule and to receive and transmit data in their assigned transmission slots. Since the new wireless MAC is only relevant for the APs and the nodes, only they have been further considered. Their hardware and software architecture will be described in the following sections.

Hardware Modules

Both components, AP and node, are realized using the same hardware platform. It is based on a standard industrial PC with an Intel Core 2 (2.1 GHz) CPU with (2 GB) DDR2 RAM. The used WLAN hardware is an Atheros AR5006X PCI based WLAN card which supports 802.11a/b/g, i. e., data rates of up to 54 Mbps can be used.

Software Modules

The software modules are being developed on an *Ubuntu Linux 10.04 LTS* using kernel version 2.6.32. The decision to use a Linux operating system (OS) is because open-source drivers for WLAN cards are almost only available for Linux. The driver module for the Atheros WLAN cards (*ath5k*) [101] serves as basis for the IsoMAC protocol extensions. The *ath5k* driver makes use of the uniform software API for WLAN interfaces *mac80211*, which controls the MAC Sublayer Management Entity (MLME) in the OS [102]. Driver modules that use *mac80211* are mainly responsible for low level hardware operations, which include packet handling and management amongst others. Both wireless modes, AP and station, are managed by the *ath5k* driver module and are set through Linux user-space processes (*hostapd*) or (*WPA supplicant*) that interact with *mac80211* [113].

Industrial real-time communication always requires minimal and bounded latency and jitter for process data transmission among connected network entities. The achievement of these goals demand optimized processing time on both sides (AP and node). That implies modifications at the lowest level to accomplish the needed performance and determinism, hence all IsoMAC related functionalities are implemented as extensions to the *ath5k* module. Moreover, specific interfaces with small processing overhead are implemented, in order to guarantee fast interaction with the IsoMAC extensions. Industrial RT Ethernet protocols have specific frame formats, hence through modifications of the Ethernet driver these frames can easily be filtered and forwarded to the IsoMAC interface. The envisaged software architecture is shown in Fig. 8.2. The IsoMAC extensions are highlighted in grey, the standard modules involved in communication are shown on the right side of the figure. These standard modules encompass various functionalities, such as the configuration of operational modes or the encapsulation of Ethernet frames. The use of uniform interfaces, like *nl80211*, increase the latency of the interprocess communication as evaluated in comparison to standard Linux interfaces [167].

In order to retain BE traffic while simultaneously adding real-time traffic, tremendous changes to the buffer handling of the *ath5k* driver module are necessary. In the TDMA based approach of IsoMAC the contention free RT traffic

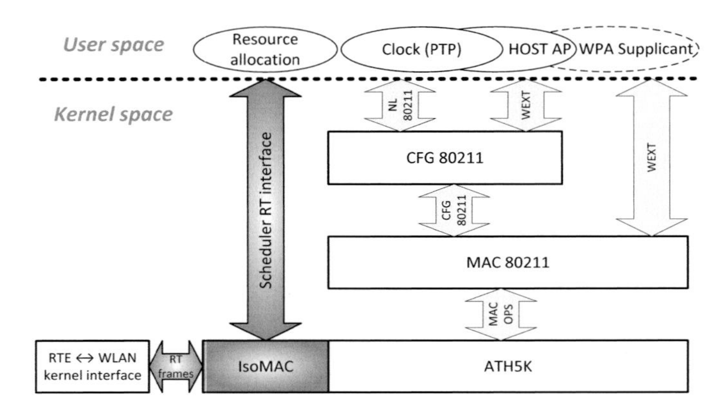

Fig. 8.2: IWN software architecture

starts right after the beacon frame as described in Sect. 5.2. Therefore any BE traffic to be sent is set into a pending state until the CFP ends. The Atheros AR5006 WLAN chipset supports different hardware queues which are controlled by the Queue Control Unit (QCU). Every queue has a different priority, as a consequence there is a mutual lock-out of these queues when trying to access the physical layer logic of the chipset. This gives a measure to control the frame flow by the driver and it enables the needed hold back of lower priority BE traffic. The beacon frame queue has the highest priority, it will not be blocked by any other frame that is to be sent by the QCU. The next lower priority queue is assigned to RT traffic.

8.2.2 IWN Access Point

The beacon transmission is controlled by the WLAN hardware, meaning that the provided packet to the corresponding queue is always send at the next Target Beacon Transmit Time (TBTT). Every action is scheduled with respect to this reference time, that holds for both hardware and software. To provide the schedule to the nodes which are associated to the AP, the beacon is updated just before the TBTT with the scheduled TX times of every real-time frame in this NUT.

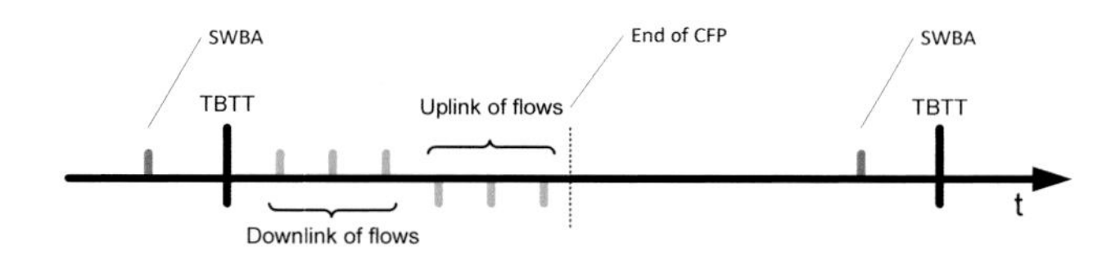

Fig. 8.3: Timeline of a NUT

This update is triggered by the Software Beacon Alert (SWBA), which generates a software interrupt. Along with the preparation of the beacon, all *downlink* frame buffers are prepared and filled with data. The hardware is configured to start transmission of the *downlink* frame queue immediately after the beacon is successfully sent out at TBTT. A sketch of the temporal dependencies of these events is given in Fig. 8.3.

8.2.3 IWN Node

The node side is more critical in terms of timing. The earliest time when the schedule for the current NUT will be available is right after the beacon has been received and processed. This shortens the available time frame for buffer preparation, which is proportional to the number of *downlink* frames and the position of the corresponding *uplink* frame of the node in the schedule. At the node side we can not resort to hardware timers starting the transmission. Every action is timed by means of the timers available in the Linux kernel API and exposed by the PC hardware. The queue has to be enabled just before the actual transmission time. Once the medium is detected to be idle, the DCF control unit (DCU) of the hardware will immediately start transmission of the *uplink* queue. In case a node does not know the current schedule, i. e., the beacon frame was not received properly, it will transmit its data during the CP with the highest priority.

8.3 Simulation Modeling

The second case study is based on a simulation model of the whole IWN including the wired network. The simulation model can be considered as an abstract representation of the real system presented in the previous section. Even though some simplifications of the system are applied, it still provides a sufficient level of detail to allow valid conclusions [4].

In this section, the conceptual core components of the simulation model are introduced. A specific focus is put on the wireless channel model which is derived from the channel characterisation in Sect. 3.2. Finally, the simulation model is validated by a comparison with the real system (cf. Sect. 8.2) and applying the correlated inspection approach of [98].

8.3.1 Conceptual Simulation Model

The presented simulation model has been implemented using the *Omnet++* discrete event simulator [170]. It is a simulation software for discrete-event simulation and mainly used to simulate communication networks. *Omnet++* uses the programming language C++ to build and describe the simulation modules. The whole simulation model is assembled by combining several simulation modules. The combination is done with the help of another high level programming language called *NED*. Moreover, the *INET framework* is used [73] which is an open source framework providing predefined modules, e. g., to simulate Ethernet or 802.11 wireless networks.

The developed conceptual simulation model of the Isochronous Wireless Network (IWN) is illustrated in Fig. 8.4 and represented as *UML* class diagram [91]. The most important modules of it are highlighted in grey and further described in the remainder of this subsection, except of the wireless channel model, which is introduced in the following Sect. 8.3.2.

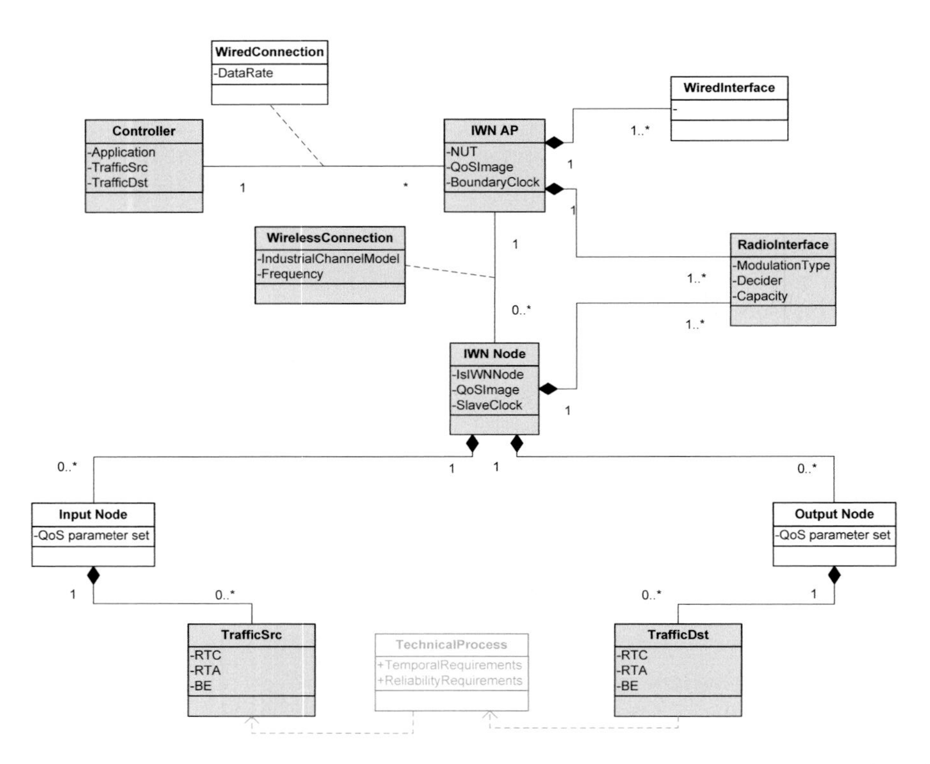

Fig. 8.4: Conceptual simulation model of the Isochronous Wireless Network (IWN)

IWN AP

The IWN Access Point (AP) is considered as the central point of control for the wireless network. It encompasses the modules for medium access control, scheduling, admission control, and clock synchronization. Furthermore, a resource monitoring is implemented which is used in the context of the admission control. It acts as a bridge between the wireless and wired communication system which is assumed to be an RTE. Moreover, it is synchronized to the wired RTE and connected to a PLC and the corresponding traffic sources and destinations.

IWN Node

The IWN node interconnects an embedded IOD through a wireless link with the AP. The IOD is represented by traffic generators real-time traffic flows. The wireless link is modelled according to the wireless channel model introduced in the next section. It consists of the medium access control, admission control, and clock synchronization. The node model only contains a reduced set of functionalities, because it is controlled by the AP.

Application and Traffic Generators

The application that controls the technical process is represented on the side as well as on the node side by traffic sources and traffic destinations, instantiated by a traffic generator which is able to generate different types of traffic. Every traffic source establishes a communication relationship to their relevant traffic destination. The traffic generators support the client/server communication model as well as the producer/consumer model. They allow to measure relevant metrics, such as the update time or latency, from an end to end perspective of the application.

8.3.2 Wireless Channel Model

The simulation of wireless communication system and the credibility of the obtained results often suffer from simplified assumptions regarding the physical layer and the wireless channel model [135]. Since wireless communication systems mainly depend on the environment and the corresponding channel characteristics, both must be sufficiently represented in the simulation model. Therefore, an evaluation based on discrete event simulation requires a precise model of the industrial wireless channel in order to deliver valid results.

One of the main differences of an industrial RF channel, in comparison to office environments, is the increased multipath effect caused by metal elements, heavy machinery and several moving objects. Multipath propagation results in a frequency selective behaviour of the channel, which makes the communication very challenging and less reliable due to an increased bit error rate. Up to now, several works exists dealing with bit error models, such as [174] and [180]. The most frequently used model is described in [180]. In this work bit error measurements have been performed with an IEEE 802.11 compliant PHY module and a thorough statistical analysis of the encountered bit errors was done. Afterwards a bit error model was derived based on the well known Gilbert/Elliot model [36]. However, the derived model is very pessimistic and leads to poor results of any investigated protocol. Furthermore, the model is based on measurements within a single environment only, and the model is lacking a representation of other possible environments in industry. Hence, a new wireless channel model is developed in this work. It is again based on measurements in a real industrial environment, but in terms of the channel transfer function $H(t, f)$. Moreover, in this approach the modulation scheme, and correction mechanisms must be also taken into account.

Packet Error Rate Model with Channel Coding

The PER model has a modular structure and consists of two different parts. First, a characterisation of the wireless channel in terms of a time and frequency varying channel transfer function $|H(t,f)|$. Second, a physical layer model which includes a decider based on the deployed modulation scheme and the error correction mechanisms.

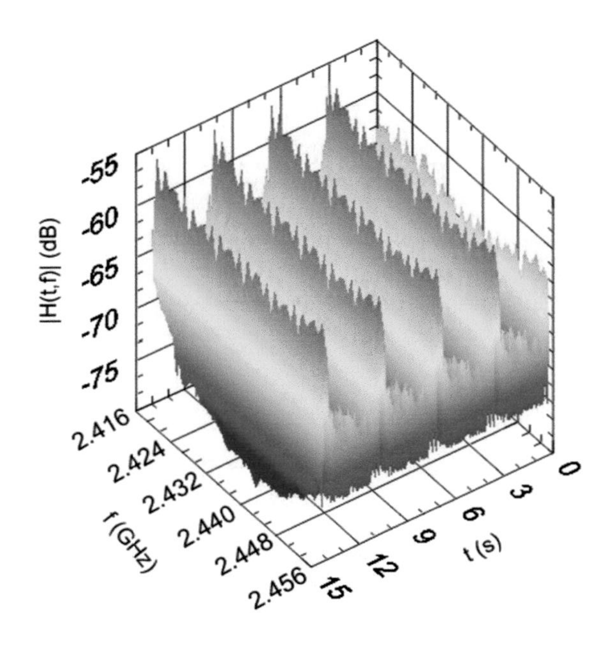

Fig. 8.5: $|H(t,f)|$ (in dB) of the wind energy plant

The wireless channel characteristics derived in Sect. 3.2 are used throughout this evaluation, since they represent the second application category. The obtained transfer function $|H(t,f)|$ is shown in Fig. 8.5 for a duration of 15 s. It can be easily seen that this channel has a periodic behaviour caused by the rotation of the system. The maximum difference of the channel gain is approx. 18 dB. Following a trace-driven approach, the different values of $|H(t,f)|$ are selected as the current channel gain G_{Ch} based on the simulation time, the temporal resolution of the trace and the selected frequency.

Such an approach has the advantage, that it has a higher similarity to the real system and requires no simplification [75]. However, well-known shortcomings of trace-driven simulations, such as representativeness, have been carefully considered in order to avoid a reduced credibility of the results [85].

The physical layer model first calculates the signal-to-noise ratio SNR (in dB) as shown in Eq. (8.1).

$$SNR = 10log(P_{Rx}) + G_{Ch} - 10log(P_{noise}) \tag{8.1}$$

where P_{Rx} is the power of the received signal, G_{Ch} the negative channel gain obtained from the real wireless channel, and P_{noise} the noise power at the receiver.

The bit error probability model which considers the SNR as input is based on the work of Miller [118] and Pei et al. [139]. It is developed for different OFDM data rates. The resulting model is validated in [139] for different modulations using results from [135] obtained with a real world testbed.

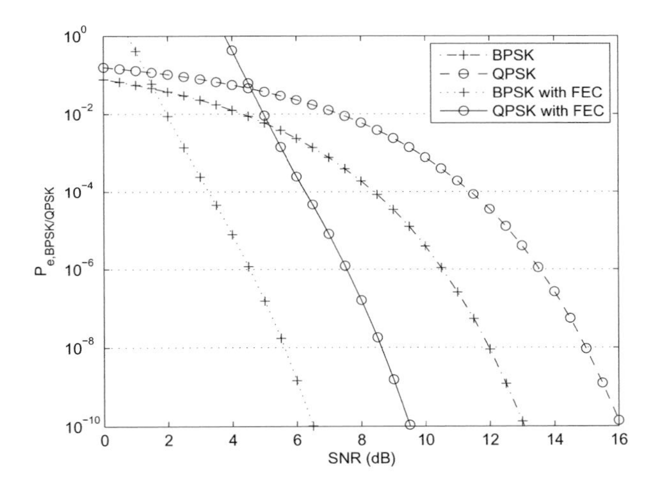

Fig. 8.6: Bit error probability with and without coding for BPSK and QPSK

The OFDM bit error probability with and without FEC is shown in Fig. 8.6 for the modulation types BPSK and QPSK, because they are most relevant for this work due to their robustness. All different options of IEEE 802.11 are shown in Sect. 2.3.

$$P_{e,BPSK} = f_{modBPSK}(SNR) \tag{8.2}$$
$$P_{e,QPSK} = f_{modQPSK}(SNR) \tag{8.3}$$

The PHY header and the MAC Protocol Data Unit (MPDU) must be differentiated due to their different modulation types. The error probability of the header is determined using BPSK with a fixed packet length $l = 24$ bit, and the packet error probability of the MPDU is determined using QPSK with packet length l in bits, depending on the payload of the application. After deriving the the bit error probability with Eq. (8.2) for BPSK and Eq. (8.3), the packet error probability P_H is determined with Eq. (8.4) for the header and P_{MPDU} with Eq. (8.5) for the MPDU.

$$P_H = 1 - (1 - P_{e,BPSK})^l \quad (8.4)$$

$$P_{MPDU} = 1 - (1 - P_{e,QPSK})^l \quad (8.5)$$

Finally, the decision whether a packet is neglected in the simulation or not is taken. Therefore, two uniformly distributed random numbers x_1 and x_2 are drawn from the interval $0 \leq x_i \leq 1$. The packet is considered to be erroneous, if either Eq. (8.6) or Eq. (8.7) holds true.

$$x_1 \leq P_H \quad (8.6)$$

$$x_2 \leq P_{MPDU} \quad (8.7)$$

Extensibility of the Wireless Channel Model

Due to the modular architecture of the presented wireless channel model, it can be easily extended using other available transfer functions of industrial channels. Several industrial channels are already characterised and available, all derived by real measurements in different industrial environments [132]. This includes channel models for:

- a real industrial production site for terminal blocks, where the measurements have been taken during normal operation, and during the setup phase of the whole production line in three different states of the plant (empty building, half of the building occupied, almost completed line). Altogether measurements at different distances are taken
- a production site within our University building which consists of several wood processing machines and a small high rack warehouse (see left hand side of the picture). Due to the present heavy machinery the whole environment can be considered as a representative for a real manufacturing site, even though it is only used for teaching purposes
- several robot and production cells in various production facilities
- both wireless channel characterisations presented in Sect. 3.2

As compared to existing approaches, the presented approach is more flexible in terms of using various characterisations of real industrial environments, since it allows to import and utilise a huge set of available channel models. Moreover, the abstraction level of the presented approach involves also modeling different modulation types and channel coding schemes for error correction. This allows a more detailed model of the investigated system. Another major advantage of the wireless model is the ability to compare real world systems with existing simulation models by using a wireless channel emulator and conducted measurements as described in the following section.

8.3.3 Model Validation

The aim of this simulation case study is to evaluate different characteristics of the proposed IWN in industrial environments. Especially, the temporal characteristics and the reliability are of interest for the system evaluation. As already identified in [137] the results of such an evaluation strongly depend on using a valid simulation model. Since the simulation model should allow to study the aforementioned aspects and derive valid conclusions being as close as possible to a real world system, the relevant metrics update time T_{IAT}, latency T_{Lat}, and packet loss rate λ_{PL} are considered in the validation.

The validation has the goal to determine whether the whole simulation model including the wireless channel model represents the behaviour of the IWN concept correctly and therefore allows to derive valid conclusions. According to [75] a simulation model can be validated with the help of 3 sources, (i) expert knowledge, (ii) measurements of the real system and (iii) theoretical analysis. Since a prototypical implementation of the system exists, it will be used for this validation. Following the discussion in [75] this is the most reliable and preferred way to validate simulation models.

However, according to [75] *"a fully validated model is a myth"*, since we are not able to validate a complex model to the full extent. Instead Jain suggests a validation with selected scenarios in order to cover the important cases. In this section, the simulation model is validated in four steps.

- Definition of a simple scenario
- Empirical measurements with the real system in accordance to the scenario
- Simulation of the same scenario
- Comparison of results from the previous two steps to validate the model

In addition to this, expert knowledge is used to check whether the results are reasonable, as recommended in [75].

The validation is based on the correlated inspection approach as described in [98]. For this approach both systems, the simulation model and the real system, are provided with identical historical input data. In addition to this, the wireless channel model is also identical, because a Wireless Channel Emulator (WCE) is used for the real system which emulates the same channel as used within the simulation model (cf. Sect. 8.3.2). Afterwards, the model output and the real system output are compared using a statistical methodology.

Validation Setup and Scenarios

Both setups for the validation, i. e., the real system and the simulation model, are shown in Fig. 8.7. One of the core parts of the real validation setup is the WCE which allows to run the real scenarios under identical environmental conditions as the simulation. This must be assured when validating wireless simulation models, but it very challenging to provide a real setup for reproducible measurements. All relevant parameters are selected to be identical for both system setups. Both parameter sets are shown in Table 8.2.

Table 8.2: Parameter sets of the validation

Parameter	Real system	Simulation
APs / Nodes	1 / 1	1 / 1
Flows per node	1 uplink, 1 downlink	1 uplink, 1 downlink
Application payload per flow (bytes)	46, 250, 500 750, 1000	46, 250, 500 750, 1000
Network update time T_{NUT} (ms)	20.48	20.48

Real system setup for validation The real setup consists of one AP and one node. The communication is bi-directional consisting of one uplink flow and one downlink flow. The application payload is varied between 46 bytes and 1000 bytes. The established traffic flows are exchanged cyclically and the whole system is configured with $T_{NUT} = 20.48$ ms. The wireless channel between transmitter and receiver is emulated by a WCE using the channel transfer function $|H(t, f)|$ derived in Sect. 3.2. Besides the node and the AP, the WCE is the core component of the setup for the real system. It operates in the RF domain and is built with coaxial cables, power splitters and programmable attenuators. This approach has the advantage of providing a reproducible setup without undesired external interferences. It also allows to emulate several other existing industrial wireless channels.

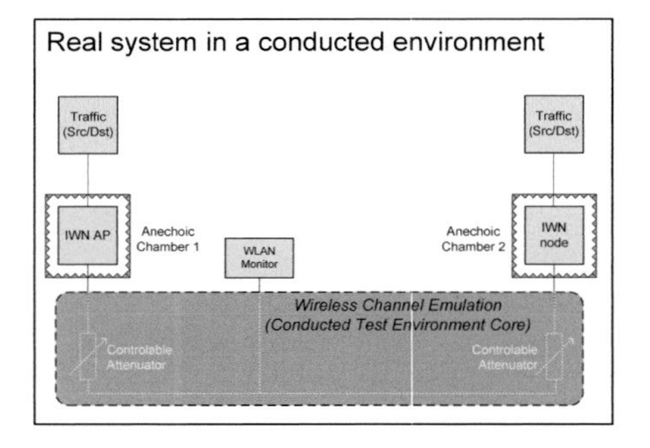

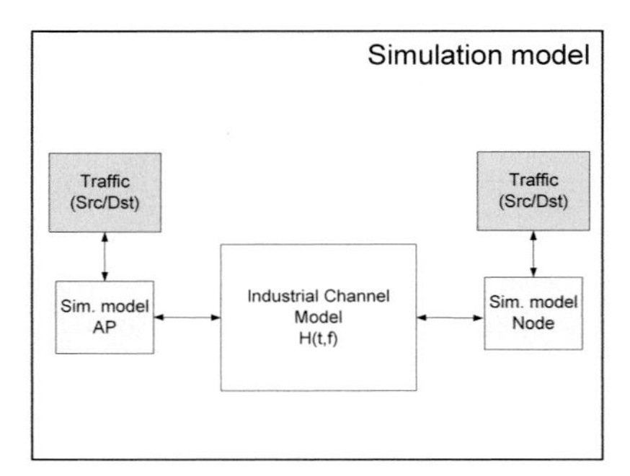

Fig. 8.7: Validation setups for the real system and the simulation

Simulation setup for validation The simulation consists of one AP and one node. The communication is bi-directional consisting of one uplink flow and one downlink flow. The application payload is varied between 46 bytes and 1000 bytes. The established traffic flows are exchanged cyclically and the whole system is configured with $T_{NUT} = 20.48$ ms. The wireless channel between transmitter and receiver is represented by the wireless channel model introduced in Sect. 8.3.2.

Validation Results and Comparison

The validation results are presented in this subsection. During the validation scenarios all metrics which are relevant for this work are recorded and compared. First of all, it is observed that all downlink transmissions within the real system have a constant offset. The offset is shown in Fig. 8.8(b) and caused by constraints of the WLAN hardware.

As shown in Fig. 8.8(a), the frame generation must be approx. 5.4 ms before the downlink frame transmission starts. Along with the preparation of the beacon, all *downlink* frame buffers must be prepared and filled with data until the SWBA time, because solely the WLAN hardware is responsible for the transmission of *downlink* frames immediately after beacon transmission, i. e., no additional downlink frames can be enqueued after the SWBA has expired. In order to not distort the validation results, this constant offset is removed from the downlink traffic of the real system by calibration.

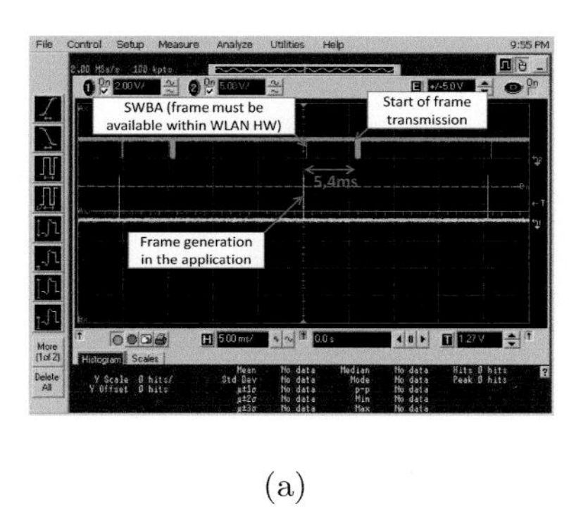

(a)

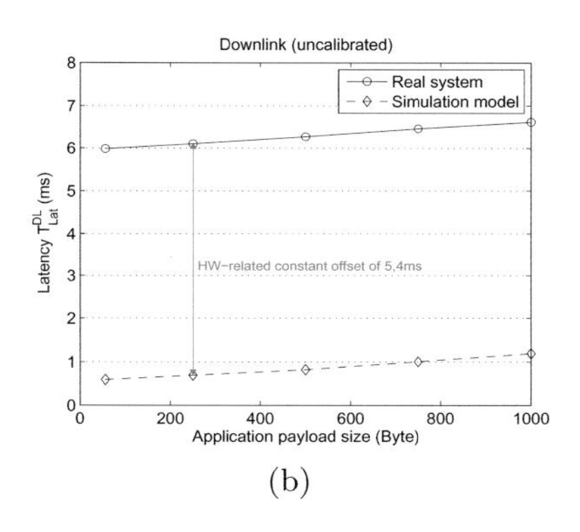

(b)

Fig. 8.8: Downlink latency offset caused by constraints of the real implementation

The validation results of the calibrated downlink latency T_{Lat}^{DL} are shown in Fig. 8.9(a). Only a minor deviation of approx. 5% can be observed for payload sizes of 500 bytes and 750 bytes, all other results are almost identical with a relative error of < 2%. The same is true for the latency T_{Lat}^{UL} uplink direction shown in Fig. 8.9(b). For the uplink case all mean values of the real system deviate by < 3% to the simulation results.

In the next scenario, the downlink update time T_{IAT}^{DL} and the uplink update time T_{IAT}^{UL} are shown in Fig. 8.8(b) and Fig. 8.10(b), respectively. In both directions the update time of the simulation model is slightly higher, by approx. 100 µs, but the relative error of both systems is < 1% in all cases. This effect is caused by the increased PLR for the simulation, described in the next section. Whenever a packet loss occurs, the next update time will be as twice as much, i. e., 40 ms, because the frame will not be retransmitted and the next frame is sent in the next NUT.

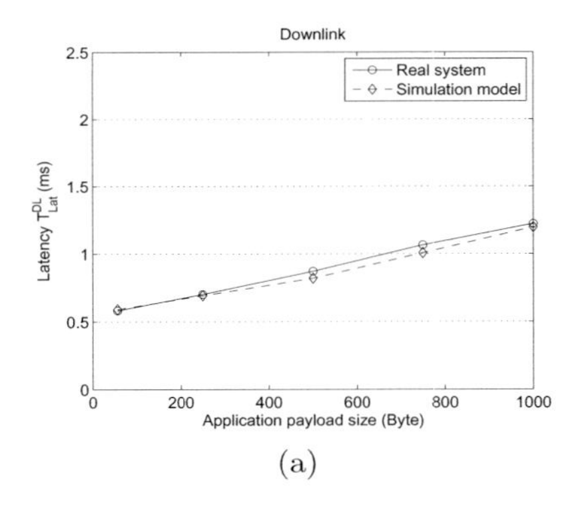

(a)

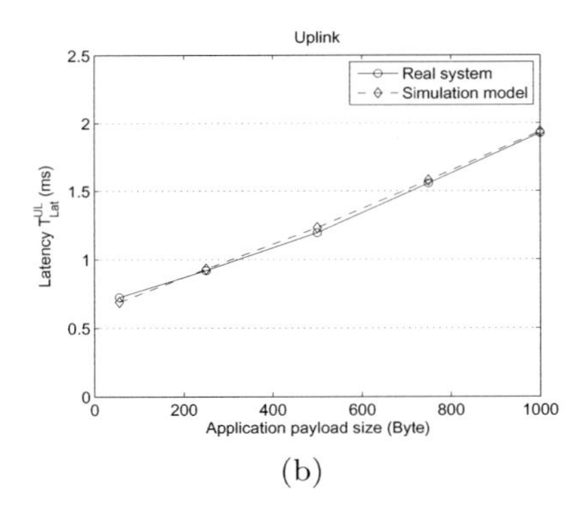

(b)

Fig. 8.9: Downlink and uplink latency T_{LAT} for different application payload sizes

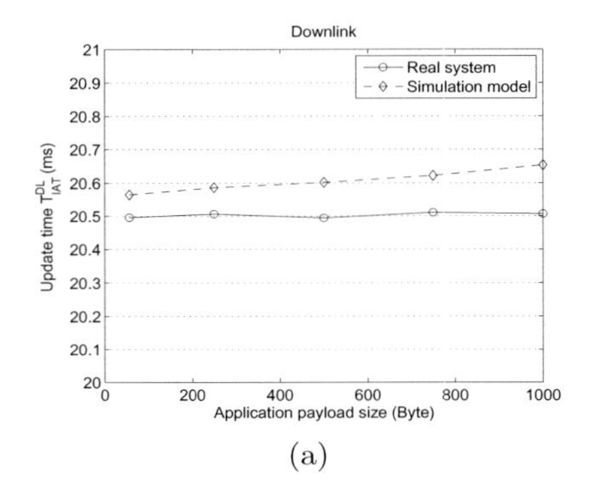

(a)

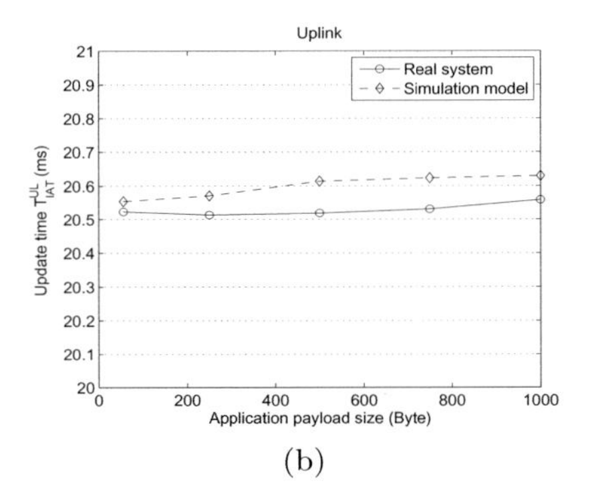

(b)

Fig. 8.10: Update time T_{IAT} in downlink and uplink direction for different application payload sizes

The PLR for the downlink direction is shown in Fig. 8.11(a) for different payload sizes. The uplink direction is shown in Fig. 8.11(b). It can be seen in both directions that the simulation results are always worse compared to the real system. This is true for all investigated payload sizes. Hence, the simulation model can be considered as more pessimistic than the real system, i. e., the reliability results of the simulation model might be even better when using a real system which is not critical.

Finally, in order to determine whether the difference of the obtained validation results is statistically significant, the first downlink scenario is further investigated and compared using the confidence interval based paired-t approach [98]. Even though, this check has been applied to both update time and latency, the necessary steps and assumptions are described only once in this section for the comparison of the update time T_{IAT}.

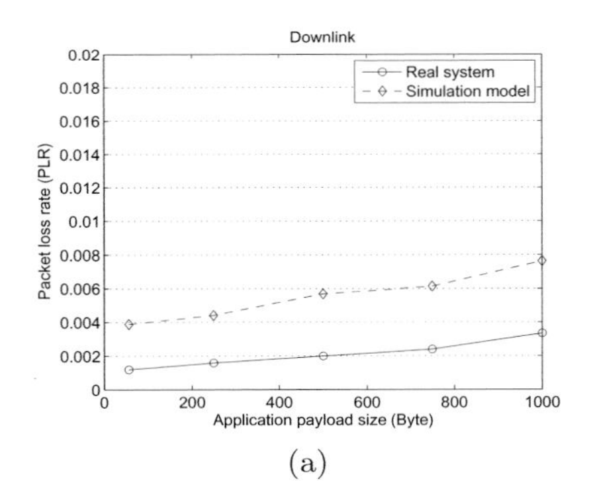

(a)

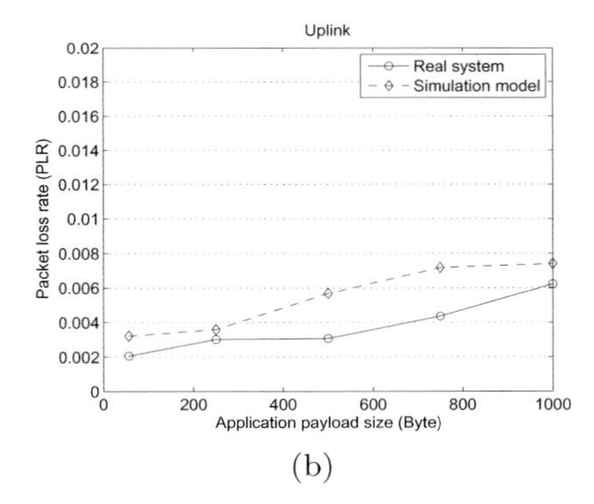

(b)

Fig. 8.11: Packet loss rate λ_{PL} for different application payload sizes

First, the approach requires to collect n sets of data from the real system and m sets of data from the simulation. Further, X_j is defined to be the average of the jth set of system data and Y_j the average of the jth set of simulation data (cf. Eq. (8.8)). The X_j's and Y_j's must be Independent and Identically Distributed (IID) random variables with $\mu_X = E(X_j)$ and $\mu_Y = E(Y_j)$. Then the model is compared with the system by constructing a confidence interval for $\zeta = \mu_X - \mu_Y$ with $l(\alpha)$ and $u(\alpha)$ being the lower and upper endpoint of the interval. Referring to [98], the average of X_j, Y_j and Z_j is given by Eq. (8.8), and the variance of Z_j is given by Eq. (8.9).

$$\overline{X}(n) = \frac{\sum_{j=0}^{n} X_j}{n} \tag{8.8}$$

$$\widehat{Var}[\overline{Z}(n)] = \frac{\sum_{j=0}^{n} [Z_j - \overline{Z}(n)]^2}{n(n-1)} \tag{8.9}$$

Letting now $Z_j = X_j - Y_j$ and $m = n = 3$, the following values are obtained for the latency using the results of the three experiments of Table 8.3:

$$\overline{Z}(3) = \overline{X}(3) - \overline{Y}(3) = 0.5921 - 0.5917 = 0.0004$$

$$\widehat{Var}[\overline{Z}(3)] = \frac{\sum_{j=0}^{3} [Z_j - \overline{Z}(3)]^2}{(3)(2)} = 3.392 \cdot 10^{\ 5}$$

and the 99% confidence interval for ζ is

$$\overline{Z}(3) \pm t_{2,0.995}\sqrt{\widehat{Var}[\overline{Z}(3)]} = 0.0004 \pm 0.0578$$

The 99% confidence interval for the latency difference is $[-0.0574, 0.0582]$ and $[-0.1386, 0.0218]$ for the update time. If $0 \in [l(\alpha)\ u(\alpha)]$, which is true for both intervals, it can be stated that the difference between the system data and the model data is not statistically significant. The results of all experiments are shown in Table 8.3.

Table 8.3: Statistical parameters of the first downlink scenario (DL/46 bytes)

	Real system		**Simulation**		**Difference**	
Experiment j	Update time (ms) U_j	Latency (calib.) (ms) X_j	Update time (ms) V_j	Latency (ms) Y_j	Diff. Update time W_j	Diff. Latency Z_j
1	20.4955	0.5834	20.5651	0.5917	-0.0695	-0.0082
2	20.5065	0.6033	20.5696	0.5918	-0.0631	0.0115
3	20.5022	0.5896	20.5449	0.5915	-0.0427	-0.0019
Mean	$\overline{U}(n)$	$\overline{X}(n)$	$\overline{V}(n)$	$\overline{Y}(n)$	$\overline{W}(n)$	$\overline{Z}(n)$
	20.5014	0.5921	20.5598	0.5917	-0.05843	0.0004

8.4 Case Study 1: Reconfigurable Manufacturing

The first case study considers the application category of reconfigurable manufacturing systems. The experiments are carried out in a real manufacturing environment. The application category can be characterised by an open environment where several wireless nodes, e. g., for maintenance or monitoring, might use the wireless system. Therefore, the IWN has to deal with uncontrollable nodes, which generate best-effort traffic. In this situation, the real-time communication should not be disturbed and the temporal requirements of real-time class 1 must be satisfied. The goal of this experiment is to show the applicability of the system in application category 1 and to evaluate it's capability to isolate real-time traffic flows from best-effort traffic.

8.4.1 Approach and Implementation of the Experiments

The presented evaluation approach consists of two different experiments. In the first experiment it is shown, that the prototypical implementation of the IWN is able to satisfy the requirements of the first application category in a realistic environment. In the second experiment, the isolation against best-effort traffic is further investigated. Here, the evaluation approach is to show the influence of the isochronous medium access control together with the resource allocation as compared to the same system with both components deactivated.

Based on the concept of our approach, the expected result is that the IWN fulfils the requirements and, at the same time, provides a sufficient isolation to best-effort traffic. First, because the medium access control and scheduling components allow a coordinated access to the medium preventing collisions and unpredictable medium access delays. Second, the system implements mechanisms to prevent best-effort traffic from being transmitted during the scheduled phase, such as setting the virtual carrier sensing and the restricted phase (cf. Sect. 5.2). Therefore, the real-time traffic flows should not experience significant influences when both components are active, whereas a major influences can be expected if both components are deactivated.

Evaluation Environment

The setting where our measurements take place is a real manufacturing environment for automated furniture manufacturing. It has a main corridor and a number of different machines disposed as work cells as well as encapsulated robot cells. The ceiling, made out of metal and acrylic, is approximately 15 m high. The shop floor area is separated by walls and thick glass from office space. A picture and the layout of the manufacturing shop floor is shown in Fig. 8.12.

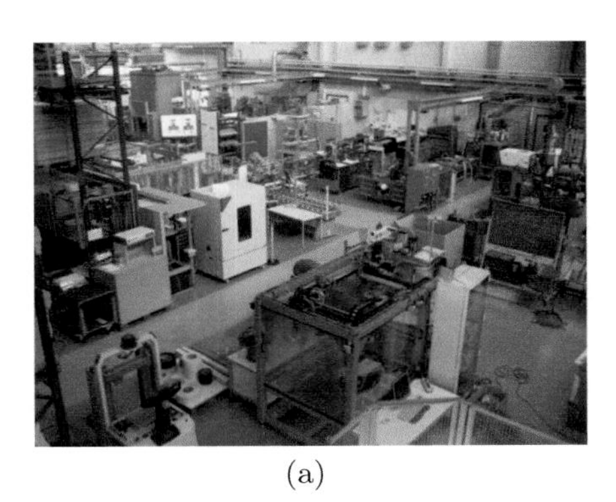

(a)

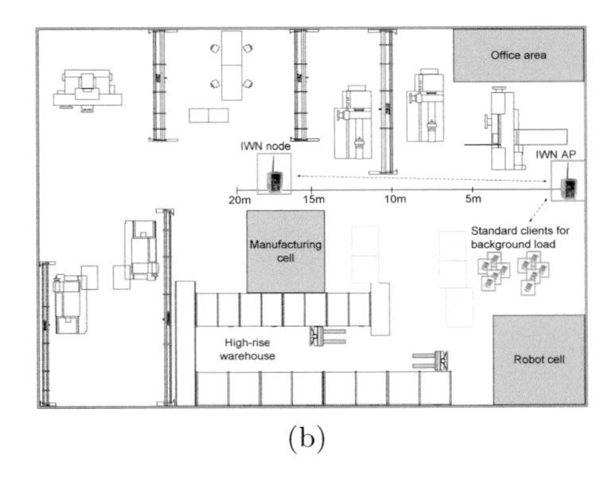

(b)

Fig. 8.12: Evaluation environment for case study 1 using the prototypical implementation (a), and groundplan including AP and node positions (b)

Implementation and Requirements

The experiments are implemented using the prototypical implementation described in Sect. 8.2. The setup consists of the IWN AP which is connected to an emulated PLC and one IWN node connected to an emulated IOD. A varying number of best-effort stations for monitoring and maintenance generate background traffic. The best-effort stations generate traffic with a constant bit rate of 0.8 Mbps per station. The sample size for the first case study is determined to be $N = 10000$ to obtain a confidence level of 95%.

The distance between the IWN AP and IWN node is approx. 15 m resembling a typical distance in reconfigurable manufacturing systems. The best-effort stations are emulated by a WLAN client emulator from Ixia which has the same distance to the AP. The parameter set of the first case study including the requirements of the first application category are listed in Table 8.4. The requirements of this application category are real-time class 1. As shown in Fig. 3.5, the response time between different flexible modules is in the range of 40 ms. Therefore, the wireless cycle time T_{NUT} is chosen to be 20.48 ms, because of the relation of the response time T_{IO} and T_{NUT} (cf. Sect. 4.3).

Table 8.4: Parameter set of the first case study

Parameter	**IWN**
APs	1
Nodes	1
Best-effort stations	$0 \ldots 8$
Real-time traffic	2 uplink, 2 downlink
flows (QoS parameter set)	T_P =20.48 ms, $d = T_P$, $OD = 2$, $l = 46 byte$,$J \leq$1 ms
Application payload per real-time flow (bytes)	46
Best-effort throughput	0.8 Mbps per station
Network update time T_{NUT} (ms)	20.48

Relevant metrics

The relevant metrics for the first case study are the latency $T_{Lat}^{UL}/T_{Lat}^{DL}$, and the update time $T_{IAT}^{UL}/T_{IAT}^{DL}$. The latency jitter is specifically relevant in the second experiment to compare the IWN with all active components to the IWN with deactivated MAC and resource allocation. The update time represents the actual sampling time of the control loop. Besides, it is used in this case study to measure the omission failures.

8.4.2 Results of the First Experiment

The evaluation results for experiment one are presented in this section, which validates the applicability of the IWN for the first application category. The main focus is put on latency and latency jitter, because both metrics are very important for the application. The boxplots in Fig. 8.13 show the latency distribution in uplink and downlink direction for a different number of best-effort stations. The number of stations varies between 0 and 8, resulting in a different bandwidth utilization of the network.

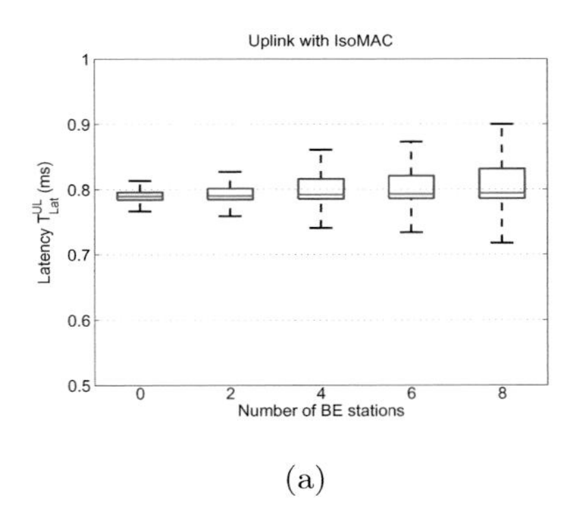

(a)

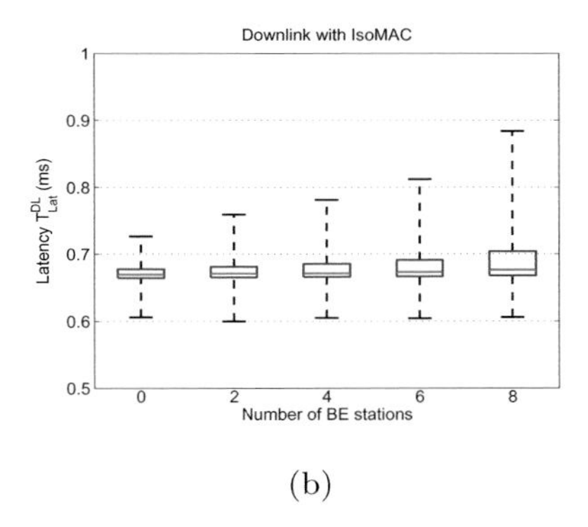

(b)

Fig. 8.13: Uplink and downlink latency T^{UL}_{Lat} and T^{DL}_{Lat} with different background loads

The presented results show that none of the real-time traffic flows is severely affected by the best-effort traffic. The mean value of the latency is almost constant and only slightly increasing. The minimum and maximum values increase for an increased number of best-effort stations, but the resulting jitter is in the range of 200 µs. The mean latency is 0.66 ms for the downlink and 0.8 ms for the uplink direction. It is shown with this experiment that the additional best-effort traffic from monitoring and maintenance stations has no negative influence on the temporal behaviour of the IWN. The implemented protection mechanisms of the medium access control and resource allocation are able to tolerate a high throughput of uncontrolled best-effort traffic without posing any severe impacts on the real-time traffic flows.

8.4.3 Results of the Second Experiment

The isolation of best-effort traffic is considered in the second experiment in order to demonstrate the relevance of the deterministic medium access control in uncontrolled environments. Therefore, the deterministic medium access as well as the resource allocation is deactivated and the experiment is executed as previously.

The latency distribution is again shown as boxplots. In Fig. 8.14(a) and Fig. 8.14(c) the results for a deactivated deterministic medium access control are shown in up- and downlink direction. Fig. 8.14(b) and Fig. 8.14(d) show again the results of experiment 1 with an adapted scaling to ease the comparison of both results. Even though the latency remains almost unchanged in the case of 0 and 2 stations, it starts to significantly increase for 4, 6, and 8 stations. When having

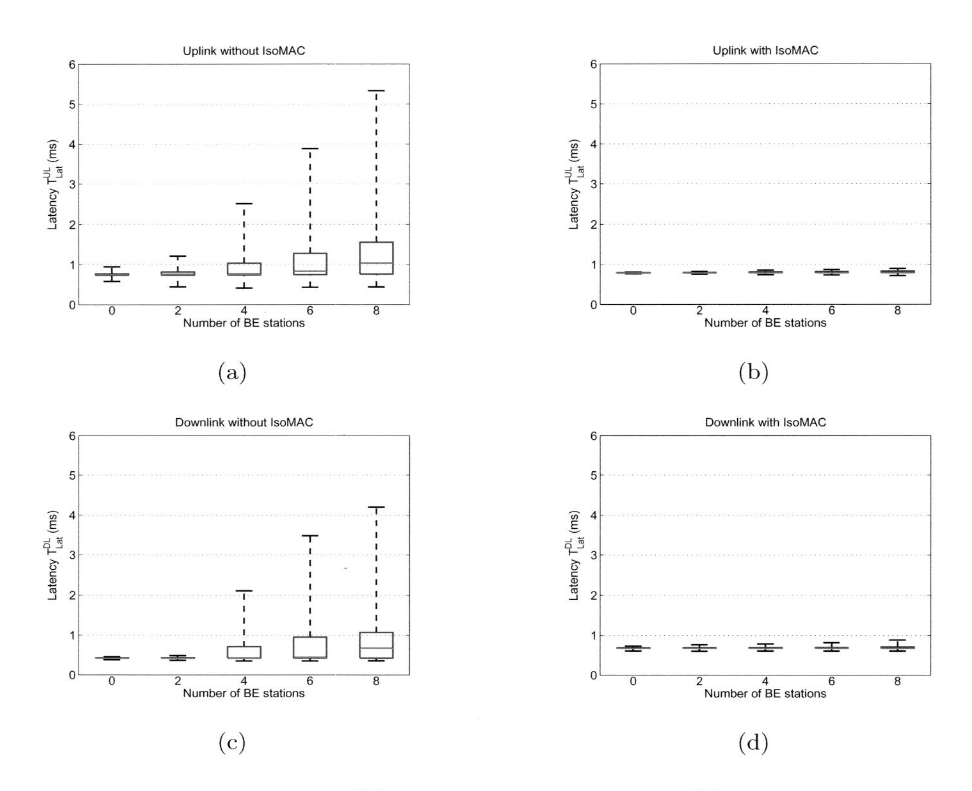

Fig. 8.14: Uplink latency T^{UL}_{Lat} and downlink latency T^{DL}_{Lat} without isochronous medium access and with isochronous medium access

8 additional station the maximum value of the uplink latency is $T^{UL}_{Lat} = 5.32\,\text{ms}$, whereas it is only 900 µs for experiment one. The new values are approx. five times higher. The same is true for the latency jitter, $J_{Lat} = 150\,\mu\text{s}$ for experiment one. In experiment two the latency jitter increases up to a value of J_{Lat} 4.82 ms (for 8 stations). Even though it is not as bad as for 8 stations, the results for all other setups are also worse than in experiment one. This shows evidently that the deterministic medium access and the resource allocation is required to satisfy temporal requirements in uncontrolled environments.

8.4.4 Summary of Case Study 1

It is shown in experiment 1 of this case study that the IWN is able to satisfy the requirements of the first application category. At the same time allowing additional best-effort stations to communicate via the IWN without a significant deterioration of the real-time traffic flows. In the second experiment, the necessity of the deterministic medium access control and the scheduling is proven. As soon as both components are deactivated, the temporal behaviour of the system significantly degrades, both in terms of latency and in terms of jitter.

8.5 Case Study 2: Wind Energy Automation

The second case study evaluates the application category of wind energy automation with the closed-loop control for the pitch motor adjustment of the blades. The wireless channel of this application category is static, as shown in the characterisation in Sect. 3.2, and only influenced by the periodic rotation with a speed of approx. 20 rpm. This application requires only a few real-time nodes, but the temporal requirements are of real-time class 2. In comparison to the first environment, almost no other best-effort nodes are present other than those for transmitting the video stream for surveillance.

8.5.1 Approach and Implementation of the Experiments

The presented evaluation approach of this case study is similar to the first one. It consists of two different experiments. In the first experiment it is shown, that the simulation model of the IWN is able to satisfy the requirements of the second application category using a realistic wireless channel model, which is designed and validated in Sect.8.3. In the second experiment two different aspects are investigated.

- First, the influence of having synchronous networks and applications. The approach here is to show the influence of deactivating the global time base component and compare the results to the first experiment, where all components are active.
- Second, the influence of different retransmission policies on the reliability of the real-time traffic flows. The results are again compared to the first experiment, but the second retransmission policy (cf. Sect. 6.4) is used, allowing more reliable real-time traffic.

Based on the concept of our approach, the expected result is again that the IWN fulfils the requirements and, at the same time, provides a sufficient capacity for the video surveillance application. First, because the medium access control and scheduling components allow a coordinated access to the medium preventing unpredictable medium access delays and the resulting latencies and jitter. Second, the system implements mechanisms to provide a global time base to allow synchronous real-time traffic flows between the wired and wireless network (cf. Sect. 5.2). Therefore, the real-time traffic flows should not experience significant influences when all components are active, whereas a major influence on the jitter is expected if the global time base is deactivated.

Evaluation Environment

The evaluation environment of the second case study is shown in Fig. 8.15. The shown nacelle is approx. 10 m long and 5 m wide. The nodes are located in the rotating part, whereas the AP is mounted in the nacelle. Within the simulation model, used for the second case study, this environment is represented as a wireless channel model of it, as described in Sect. 8.3.2.

(a)

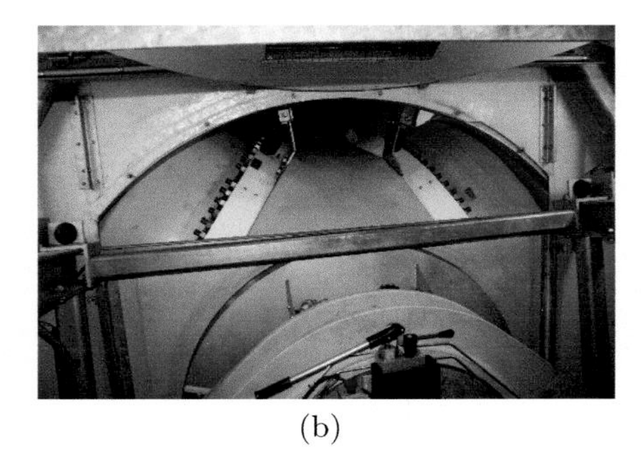

(b)

Fig. 8.15: Environment of the second case study represented as wireless channel model in the simulation

Implementation and Requirements

Both experiments are implemented using the simulation model described in Sect. 8.3. The simulation scenario consists of the IWN AP which is connected via a synchronous network to the PLC. Five IWN nodes, which include an IOD, are connected to the AP via the channel model . In addition to the real-time traffic flows, a video stream for surveillance of the blades is transmitted. The compressed video stream has an average throughput of 1.2 Mbps, as commonly for this type of traffic (cf. Sect. 3.1.1). The sample size for the second case study is determined to be $N \approx 40000$ to obtain a confidence level of 99%.

The distance between the IWN AP and IWN nodes is approx. 5-10 m resembling a typical distance found in the nacelle of a wind turbine. Video traffic is transmitted via a separate best-effort station, which has the same distance to the AP. The parameter set of the first case study including the requirements of the first application category are listed in Table 8.5. The requirements of this application category are real-time class 2, with required latency $T_{LAT} \leq 10$ ms and a maximum latency jitter of $J_{Lat} \leq 100$ μs. The wireless cycle time T_{NUT} is chosen to be 4 ms, because of the relation of the response time T_{IO} and T_{NUT} (cf. Sect. 4.3).

Table 8.5: Parameter set of the second case study

Parameter	IWN
APs	1
Nodes	5
Best-effort stations	1
Real-time traffic	5 uplink, 5 downlink
flows (QoS parameter set)	T_P =4 ms, $d = T_P$, $OD = 2$, l =46 bytes, $J \leq$100 µs
Application payload per real-time flow (bytes)	46 bytes
Best-effort throughput	1.2 Mbps per station
Network update time T_{NUT} (ms)	4

Relevant metrics

Relevant metrics for the second case study are the latency $T_{Lat}^{UL}/T_{Lat}^{DL}$, the latency jitter, the update time $T_{IAT}^{UL}/T_{IAT}^{DL}$. The latency contributes to the overall response time T_{IO} in both directions. It must stay below the given requirement, the same is true for the jitter J_{Lat}. Both are directly influencing the control loop and the technical process. The update time represents the actual sampling time of the control loop. Besides, it is used in this case study to measure the omission failures. The metrics are defined in Sect. 4.1.2. Moreover, the achievable throughput for the video stream is recorded and investigated.

8.5.2 Results of the First Experiment

In the first experiment the applicability of the IWN is validated. The main focus is put on latency and latency jitter, because they are most important in closed-loop control system. Both histograms in Fig. 8.16 show the latency value for all nodes in uplink and downlink direction. Looking at the results qualitatively, it can be seen that none of the flows has a high deviation from its mode. The mean latency range from 0.537 ms to 0.913 ms for the downlink direction and from 1.001 ms to 1.384 ms for the uplink direction. These values depend on the synchronization point, which has been selected to be at the beginning of the beacon frame.

The update times, shown in Fig. 8.17, directly show the number of frame omissions. The histograms are for node 4, the node in the system with the worst update times T_{IAT}. The red line on the right side of both diagrams highlights the upper bound, all update times must stay below this boundary. It can be seen that the number of consecutive omissions are always on the left hand side of the red line, i. e., OD_{max} is never exceeded.

The main results of this experiment are summarized in Table 8.6 for the temporal metrics. The table shows the update time T_{IAT}, the rate jitter J_{IAT}. The

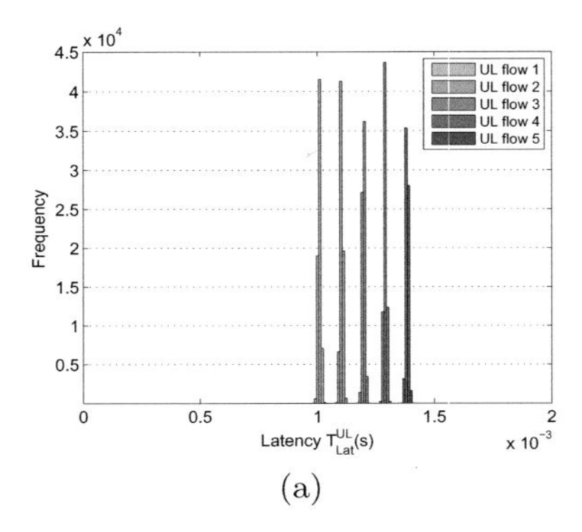

(a)

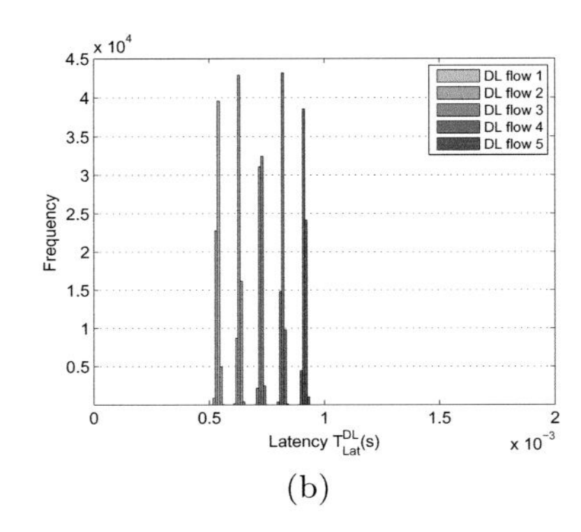

(b)

Fig. 8.16: Histograms of the uplink latency T_{Lat}^{UL} and the downlink latency T_{Lat}^{DL} for all real-time flows

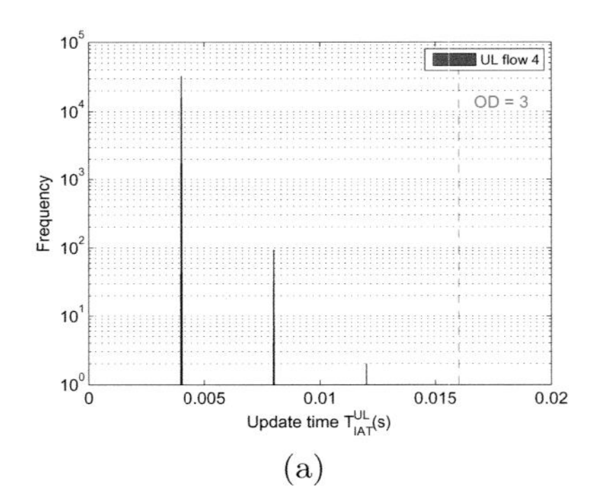

(a)

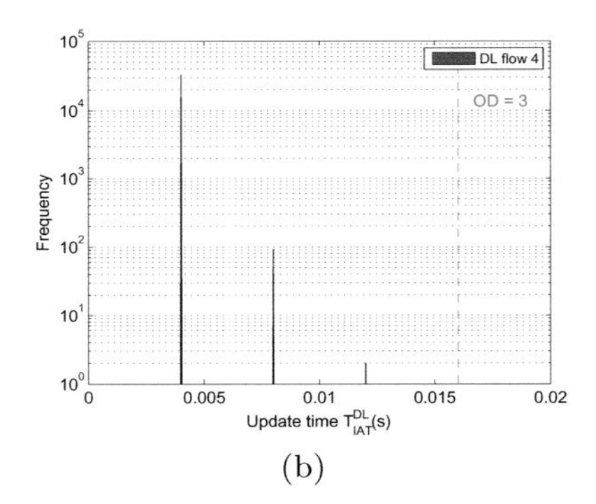

(b)

Fig. 8.17: Histograms of the update time T_{IAT}^{UL} and T_{IAT}^{DL} for node 4

update time represents the control cycle of the process. The rate jitter is directly influenced by the occurring omissions, i. e., in case of the first omission failure the frame is not retransmitted. Therefore, the update time increases by 4 ms, resulting in an increased jitter. The latency as well as the latency jitter is the important metric for the control loop. The minimal latency in downlink direction is 0.5371 ms and experienced by node 1, whereas node 5 has the highest latency of . The same is valid for the uplink direction. The latency of node 1 is 1.0081 ms, whereas node 5 has the highest latency of 1.3841 ms. The latency jitter values for all nodes stay below 55 µs. These values are able to satisfy the requirements of the second application category. Moreover, the video surveillance throughput of the best-effort station is determined to be 1.225 Mbps. However, the best-effort station experiences a higher packet loss rate and an increased number of omission, both is not critical for the surveillance application.

Table 8.6: Temporal metrics for all real-time traffic flows in experiment 1

	Node 1		Node 2		Node 3		Node 4		Node 5	
	Mean (ms)	J_x (ms)	Mean (ms)	J_x (ms)	Mean (ms)	J_x (ms)	Mean (ms)	J_x (ms)	Mean (ms)	J_x (ms)
T_{IAT}^{UL}	4.0112	4.0510	4.0113	4.0521	4.0093	4.0580	4.0111	8.0418	4.0120	4.0522
T_{IAT}^{DL}	4.0116	4.0498	4.0123	8.0269	4.0103	8.0288	4.0119	8.0485	4.0114	4.0560
T_{Lat}^{UL}	1.0081	0.0480	1.1021	0.0458	1.1961	0.0463	1.2901	0.0507	1.3841	0.0463
T_{Lat}^{DL}	0.5371	0.0483	0.6312	0.0520	0.7252	0.0451	0.8192	0.0507	0.9132	0.0430

The reliability results are shown in Table 8.7 with respect to the packet loss rate λ_{PL} and the maximum omission degree of flow OD_{max}. The results are presented for every node in both directions, uplink (UL) and downlink (DL), including the video stream. The worst $\lambda_{PL} = 0.31\,\%$ is experienced by node 2 in downlink direction and node 5 in uplink direction with $\lambda_{PL} = 0.30\,\%$. However, both nodes have a bounded omission degree of 2, and the application is not influenced. It can be noticed that non of the flows exceeds the maximum omission degree of 2 required by the application.

Table 8.7: Reliability metrics for all real-time traffic flows in experiment 1

		Node 1	Node 2	Node 3	Node 4	Node 5	Video
λ_{PL} (%)	UL	0.2788	0.2818	0.2321	0.2759	0.3007	0.6853
	DL	0.2905	0.3066	0.2555	0.2964	0.2847	–
OD_{max}	UL	1	1	1	2	2	5
	DL	1	2	2	2	1	–

8.5.3 Results of the Second Experiment

The second experiment is further divided into two scenarios. In the previous experiment, the global time base was active and enabled a synchronous wired and wireless network and application. The consequences of deactivating the global time base component are investigated first. It is based on the metric latency and the results are shown for the uplink and downlink direction, but the behaviour of the traffic flows is almost identical.

In the first experiment, retransmission policy one (cf. Sect. 6.4) is used to optimize the latency jitter J_{Lat}, because this is most important for the application category. However, to show its impact on the reliability of the real-time traffic flows as well as on the resulting jitter, retransmission policy two is examined in the second scenario. The obtained results are again compared to the first experiment in terms of the reliability metrics λ_{PL} and OD_{max}.

Deactivated Global Time Base

The deactivated global time base leads to an asynchronous behaviour of the wired and wireless network. The consequences can be seen in Fig. 8.18.

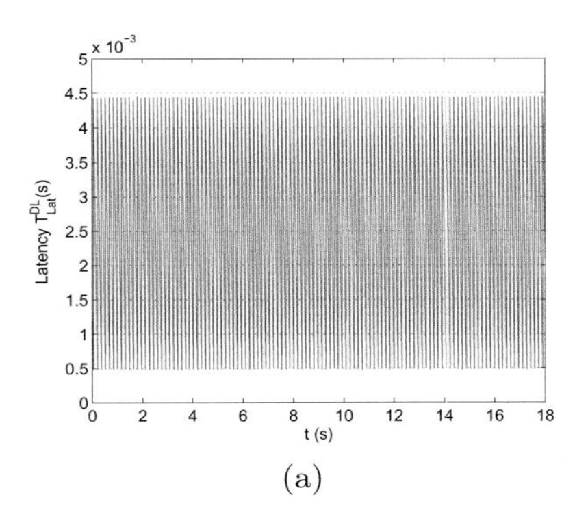

(a)

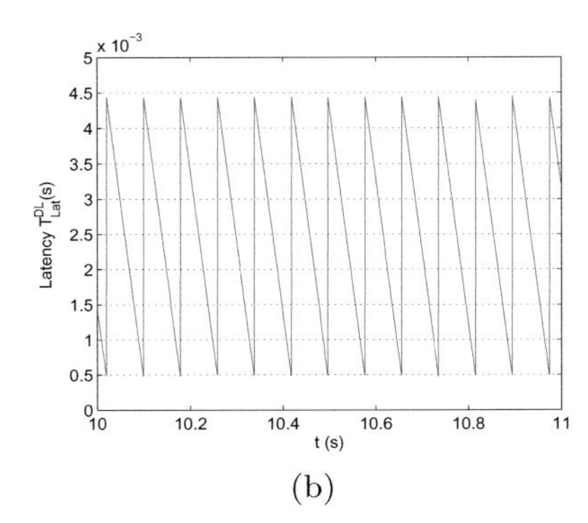

(b)

Fig. 8.18: Downlink latency T^{DL}_{Lat} of node 1 with deactivated global time base

Taking a closer look at the characteristics of the signal in Fig. 8.18(b), it can be seen that it exhibits a sawtooth pattern. Since the wireless and wired system are not synchronized, T_{NUT} and the cycle time of the wired system are slowly drifting away from the beginning of the slot time, which is allocated to the flow. This continues until a frame is kept so long in the buffer that the new packet arrives before the old one could be sent. If only the newest data is important, the old packet is dropped and the new one is sent. At that point the transmission time drops back to the minimum with one omission failure.

The resulting jitter can be approx. calculated from the boxplots in Fig. 8.19. It is almost constant for all nodes. The maximum latency value $max(T_{Lat}) \approx 4.5\,\text{ms}$, the minimum value $min(T_{Lat}) \approx 0.45\,\text{ms}$. Thus, the resulting jitter is $J_{Lat} = 4.05\,\text{ms}$.

Retransmission Policies

The result of applying retransmission policy two, i. e., frames are always retransmitted. In such a case it is more important for the application to receive a frame, and the degraded jitter is not important. The results are illustrated in Fig. 8.20(b) as histogram for the uplink latency of all 5 nodes and as update time for uplink flow 1.

Due to the retransmission the update time and the latency is significantly widened. The main results of this scenario are summarized in Table 8.9 for the temporal characteristics and in Table 8.9 for the reliability characteristics. The temporal characteristics, i. e., mainly the jitter, is deteriorated, because of the retransmissions. For instance, the jitter values in the first experiment are all in the

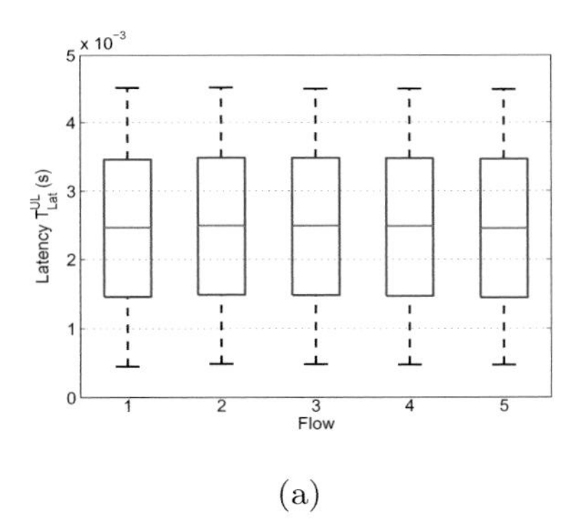

(a)

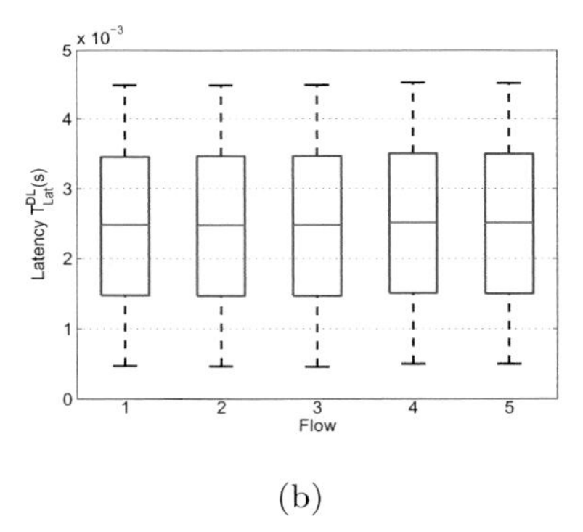

(b)

Fig. 8.19: Boxplots of the uplink T^{UL}_{Lat} latency and the downlink latency T^{DL}_{Lat} for all real-time flows

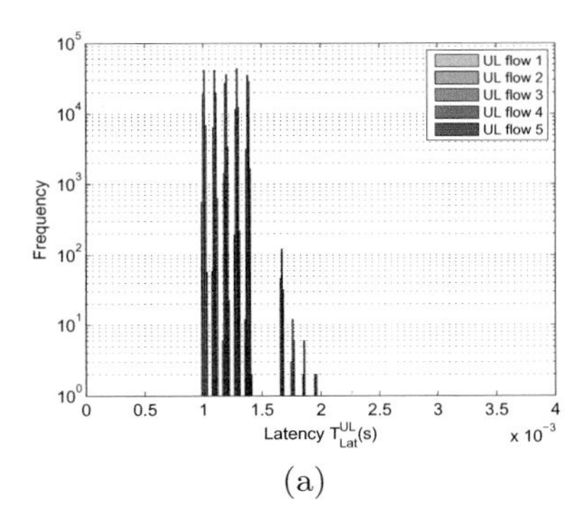

(a)

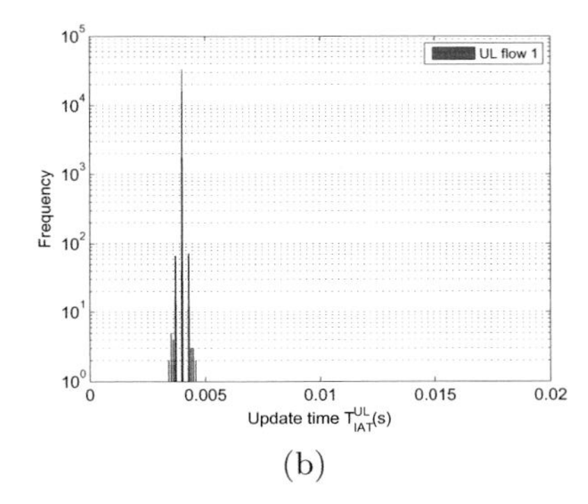

(b)

Fig. 8.20: Histograms of T^{UL}_{Lat} for all 5 nodes and the update time T^{UL}_{IAT} for flow 1

range of some ten µs. In this scenario the smallest jitter value is experienced by node 5 with 595.1 µs.

Table 8.8: Temporal metrics for all real-time traffic flows in experiment 2

	Node 1		Node 2		Node 3		Node 4		Node 5	
	Mean (ms)	J_x (ms)	Mean (ms)	J_x (ms)	Mean (ms)	J_x (ms)	Mean (ms)	J_x (ms)	Mean (ms)	J_x (ms)
T^{UL}_{IAT}	4.000	1.1587	4.000	1.3268	4.000	1.8926	4.000	1.5160	4.000	1.9015
T^{DL}_{IAT}	4.000	2.6369	4.000	2.8544	4.001	5.5076	4.000	2.8098	4.001	5.1426
T^{UL}_{Lat}	1.0103	0.9751	1.1038	0.7389	1.1975	0.9669	1.2913	0.6844	1.3849	0.5951
T^{DL}_{Lat}	0.5402	1.3420	0.6338	1.4517	0.7278	1.5355	0.8214	1.4268	0.9152	1.1683

As the results in Table 8.9 show, the reliability of wireless transmission is significantly increased. In this scenario only three nodes out of ten suffer under lost frames and the resulting very small λ_{PL}. The frame omissions have also decreased to only one.

Table 8.9: Reliability metrics for all real-time traffic flows in experiment 2

		Node 1	Node 2	Node 3	Node 4	Node 5
$\lambda_{\mathbf{PL}}$ (%)	UL	0	0	0	0	0
	DL	0.00146	0	0.00145	0.00149	0
$\mathbf{OD_{max}}$	UL	0	0	0	0	0
	DL	0	0	1	1	0

8.5.4 Summary of Case Study 2

It is shown in case study 2 that the IWN is able to support the requirements of application category two. In the second experiment, it is validated that the deactivation of the global time base leads to an increased latency and a jitter in the same order of magnitude. Finally, different retransmission policies ensure an optimized operation mode depending on the application. The adaptive handling of retransmissions allow a trade-off between a required bounded jitter as in experiment one. In this case the wireless communication is less reliable, but the application operates as expected as long as the given OD_{max} is not exceeded. The main results of the second case study are summarized in Table 8.6 for the reliability metrics and in Table 8.7 .

8.6 Concluding remarks

Based on two different application categories, it is shown in this evaluation that the IWN is able to satisfy the requirements of both considered applications.

In the first application category the IWN is able to satisfy the requirements of real-time class 1. In addition to this, the system provides an isolation for best-effort traffic originating from uncontrollable stations in the network. Thus, the real-time traffic flows are not significantly influenced in such a situation, even if the bandwidth utilization of the uncontrolled stations is very high. In the experiment, the throughput of best-effort traffic reached approx. 7 Mbps.

The requirements of the second application category are also satisfied by the IWN. The system provides a sufficient capacity for the video surveillance application without influencing the real-time traffic flows. It is shown in the evaluation that the IWN is able to provide latency guarantees $T_{Lat} \leq 10\,\text{ms}$ and a latency jitter $J_{Lat} \leq 100\,\mu\text{s}$.

9

Conclusion and Future Research Directions

9.1 Conclusion

The realisation of a highly flexible, complex, labour divided, and geographically distributed production is one of the main topics of the German initiative "Industry 4.0". In order to achieve this goal, new ways of interconnecting industrial facilities are necessary, for instance by using innovative wireless technologies. In addition to this, the efficiency and flexibility of control systems can be increased significantly. This is mainly due to applications which consist either of moving components, such as rotating machine parts, or require a high degree of mobility.

In this thesis, an isochronous wireless network for industrial control applications has been designed providing guaranteed latencies and jitter. The main challenges in this area are a non-deterministic medium access of existing systems, the uncontrolled, shared wireless medium and its limited capacity, as well as the asynchronous behaviour of wireless and wired communication systems degrading the temporal behaviour of such hybrid systems.

In order to address these challenges the presented solution approach of an isochronous wireless network for industrial automation consists of three conceptual components, the deterministic medium access control, the resource allocation, and a global time base.

A deterministic medium access control, based on a TDMA-scheme, is responsible to provide mechanisms for avoiding unpredictable latencies, introduced by uncoordinated distributed MAC protocols. It is designed with the objective to be efficient for cyclic process data frames with small payloads. It provides a sufficient capacity for high throughput applications and is interoperable with existing wireless nodes and their operation does not negatively influence the system.

The resource allocation is responsible for allocating the required resources for real-time traffic flows. It considers currently admitted traffic flows of the AP and the available capacity of the channel. It also guarantees that the application requirements can be met by all admitted flows. During the admission procedure, application defined priorities are used to be able to reject flows with a lower priority. In order to optimize the communication jitter in the presence of frame errors, an adaptive retransmission handling following different policies depending on the application requirements is deployed.

The provision of a global time base ensures that all system components including the wired nodes rely on the same global time base. Thus a significant deterioration of the temporal system behaviour by asynchronous communication processes has been avoided. Besides guaranteeing a synchronous communication, the TDMA-based medium access control requires synchronized wireless nodes with a sufficient accuracy. This has been achieved by an efficient clock synchronization protocol without additional overhead using beacon frames.

A detailed characterisation of the industrial environment has been conducted in terms of the industrial wireless channel and the analysis of two real industrial manufacturing systems. The results of the wireless channel characterisation were used to design a flexible model of real industrial channels for discrete event simulation. The investigation of the manufacturing system was the basis for deriving application requirements from a real system.

The described components allow the isochronous wireless network to fulfil the tight timing requirements of the considered NCS applications. The presented approach is able to provide guarantees for real-time traffic flows, even in the presence of uncontrollable best-effort nodes used for other purposes, such as monitoring and maintenance. It provides also support for high throughput video surveillance applications. Due to the provision of a global time base, the system can be seamlessly integrated into existing RTE networks without compromising the temporal behaviour of the whole system.

An implementation prototype of the proposed wireless system and a simulation case study have been used for the evaluation of the solution approach in two case studies. The prototypical implementation is used for the evaluation in a real manufacturing environment as well as for the validation of the simulation model. Due to given scalability constraints of the prototype, a second case study based on a realistic simulation model is conducted. A realistic channel model for the simulation, implemented based on the channel characterisation, allows more realistic simulation results.

Two different case studies have been used to evaluate the solution approach. The evaluation results show that the latency and jitter requirements of the given application of an NCS can be achieved with the presented solution approach as long as all components are active. Thus the solution approach allows a deployment within NCSs with the given requirements. As soon as components are deactivated, the behaviour of the network is significantly degraded and the requirements cannot be satisfied any more.

9.2 Future Research Directions

Even though the presented solution approach can be deployed in NCS having requirements of real-time class 2, some open research problems have been identified during the course of this work and should be addressed in future research.

To further increase the efficiency of the proposed deterministic medium access control, especially in terms of frames with a small payload, a wireless dynamic frame packing could be designed. The main idea would be to combine several single downlink frames in one big downlink frame sent as multicast to all real-time nodes. However, due to an increased frame error probability of large frames, such an approach must be carefully investigated with respect to reliability and jitter resulting from retransmissions.

In order to deal with the problem of the half duplex wireless channel and the resulting limited capacity, as compared to wired networks, a promising solution could be to use frequency division duplex (FDD) and establish a full duplex communication link. This would lead to a better temporal behaviour, because it would allow to send in both directions, uplink and downlink, at the same time. However, the resulting increased spectrum usage by such a system poses another challenge due to the crowded ISM-bands and a lack of solely available frequencies for industrial applications.

A more realistic modelling of the application in the simulation would allow to better investigate the influences on the stability of the control loop. Such a system could be achieved by means of a co-simulation approach. For instance, the whole closed-loop control system is modeled with Matlab/Simulink, as commonly used in this domain. This model provides an interface to a discrete event simulation tool, such as Omnet++, which is used to simulate the temporal behaviour of the network.

Finally, looking at one of the latest research calls of the German Federal Ministry of Education and Research (BMBF) entitled "Reliable wireless communication for factory automation", it can be noticed that also the BMBF recognized there is still further research needed. The BMBF has identified that there are limitations in today's wired implementations, which can be circumvented by using wireless technologies. The wireless systems have to fulfil at least the requirements of the wired systems like CAN, Profibus, or Profinet with respect to transmission rates and latencies. Solutions based on existing wireless systems, such as 802.15.1, 802.11, or WirelessHART, are not able to fulfil these requirements.

Thus, research on completely new wireless technologies should be conducted within the funded projects, to be finally able to satisfy the requirements of real-time class 3 for motion control loops. The requirements for the projects are derived from existing RTE solutions, i. e., a very low response time of $\leq 1\,\text{ms}$ and a jitter of $\leq 20\,\mu\text{s}$) must be achieved. At the same time a high reliability of the communication with a PER $< 10^{-9}$ and a high density of nodes must be guaranteed. The supported data rate can be rather low, because of the small payload sizes.

References

1. Alves, M., Tovar, E., Vasques, F., Hammer, G., Rother, K.: Real-time communications over hybrid wired/wireless PROFIBUS-based networks. In: Real-Time Systems, 2002. Proceedings. 14th Euromicro Conference on. pp. 142 – 151 (2002)
2. Arora, A., Yoon, S.G., Choi, Y.J., Bahk, S.: Adaptive TXOP Allocation Based on Channel Conditions and Traffic Requirements in IEEE 802.11e Networks. Vehicular Technology, IEEE Transactions on 59(3), 1087–1099 (March 2010)
3. Bang, Y., Han, J., Lee, K., Yoon, J., Joung, J., Yang, S., Rhee, J.K.K.: Wireless network synchronization for multichannel multimedia services. In: Proc. 11th International Conference on Advanced Communication Technology ICACT 2009. vol. 02, pp. 1073–1077 (15–18 Feb 2009)
4. Banks, J., Carson, J.S.: Discrete-event system simulation. Pearson Education India, 5th edn. (2005)
5. Bartolomeu, P., Fonseca, J., Vasques, F.: Implementing the wireless FTT protocol: A feasibility analysis. In: Emerging Technologies and Factory Automation (ETFA), 2010 IEEE Conference on. pp. 1 –10 (Sept 2010)
6. Bauer, M., May, G., Jain, V.: A wireless gateway approach enabling industrial real-time communication on the field level of factory automation. In: Emerging Technology and Factory Automation (ETFA), 2014 IEEE. pp. 1–8 (Sept 2014)
7. Benra, J., Halang, W.: Software-Entwicklung für Echtzeitsysteme. Springer Berlin / Heidelberg (2009)
8. Bianchi, G.: Performance analysis of the IEEE 802.11 distributed coordination function. IEEE Journal on Selected Areas in Communications 18(3), 535–547 (2000)
9. Block, D., Trsek, H., Meier, U.: Real-Time Characterization of Fast-Varying Industrial Wireless Channels. In: RADCOM 2013 - Radar, Communication and Measurement. Hamburg, Germany (Apr 2013)
10. Broomhead, T., Ridoux, J., Veitch, D.: Counter Availability and Characteristics for Feed-forward Based Synchronization. In: International IEEE Symposium on Precision Clock Synchronization for Measurement, Control and Communication (ISPCS'09). pp. 29–34. Brescia, Italy (Oct 2009)
11. Bundesverband Erneuerbare Energie: Stromversorgung 2020 - Wege in eine moderne Energiewirtschaft. online `http://www.bee-ev.de/_downloads/publikationen/studien/2009/090128_BEE-Branchenprognose_Stromversorgung2020.pdf` (Januar 2009), accessed: 03.09.2014
12. Buttazzo, G.C.: Hard Real-time Computing Systems: Predictable Scheduling Algorithms And Applications (Real-Time Systems Series), vol. 24. Springer US (2011)

13. Caida: Center for Applied Internet Data Analysis. online http://www.caida.org (2014), accessed: 03.10.2014
14. Callado, A., Kamienski, C., Szabo, G., Gero, B., Kelner, J., Fernandes, S., Sadok, D.: A Survey on Internet Traffic Identification. IEEE Communications Surveys & Tutorials 11(3), 37–52 (2009)
15. Cao, J., Chen, A., Widjaja, I., Zhou, N.: Online Identification of Applications Using Statistical Behavior Analysis. In: Proc. IEEE Global Telecommunications Conf. IEEE GLOBECOM 2008. pp. 1–6 (2008)
16. Casetti, C., Chiasserini, C.F., Fiore, M., Garetto, M.: Notes on the inefficiency of 802.11e HCCA. In: IEEE Vehicular Technology Conference VTC2005 (September 2005)
17. Cena, G., Bertolotti, I., Scanzio, S., Valenzano, A., Zunino, C.: Synchronize your watches: Part I: General-purpose solutions for distributed real-time control. Industrial Electronics Magazine, IEEE 7(1), 18–29 (2013)
18. Cena, G., Scanzio, S., Valenzano, A., Zunino, C.: Implementation and Evaluation of the Reference Broadcast Infrastructure Synchronization Protocol. Industrial Informatics, IEEE Transactions on PP(99), 1–1 (2015)
19. Cena, G., Seno, L., Valenzano, A., Zunino, C.: On the Performance of IEEE 802.11e Wireless Infrastructures for Soft-Real-Time Industrial Applications. Industrial Informatics, IEEE Transactions on 6(3), 425 –437 (August 2010)
20. Cena, G., Valenzano, A., Vitturi, S.: Wireless Extensions of Wired Industrial Communications Networks. In: 5th IEEE International Conference on Industrial Informatics, 2007. vol. 1, pp. 273 – 278 (June 2007)
21. Cena, G., Valenzano, A., Vitturi, S.: Hybrid wired/wireless networks for real-time communications. Industrial Electronics Magazine, IEEE 2(1), 8 –20 (March 2008)
22. Cena, G., Scanzio, S., Valenzano, A., Zunino, C.: Dynamic duplicate deferral techniques for redundant Wi-Fi networks. In: Emerging Technology and Factory Automation (ETFA), 2014 IEEE. pp. 1–8 (Sept 2014)
23. Chamaken, A., Litz, L.: Joint Design of Control and Communication in Wireless Networked Control Systems. at - Automatisierungstechnik 58(4), 192 – 205 (April 2010)
24. Cicconetti, C., Lenzini, L., Mingozzi, E., Stea, G.: Design and performance analysis of the Real-Time HCCA scheduler for IEEE 802.11e WLANs. IEEE Computer Networks (2007)
25. Cooklev, T., Eidson, J., Pakdaman, A.: An Implementation of IEEE 1588 Over IEEE 802.11b for Synchronization of Wireless Local Area Network Nodes. IEEE Transactions on Instrumentation and Measurement 56(5), 1632–1639 (October 2007)
26. Costa, R., Portugal, P., Vasques, F., Moraes, R.: Comparing RT-WiFi and HCCA approaches to Handle Real-Time Traffic in Open Communication Environments. In: 17th IEEE International Conference on Emerging Technologies and Factory Automation. Krakow, Poland (Sep 2012)
27. Cullen, P., Fannin, P., Molina, A.: Wide-band measurement and analysis techniques for the mobile radio channel. Vehicular Technology, IEEE Transactions on 42(4), 589–603 (Nov 1993)
28. Deppe, J., Trsek, H., Jasperneite, J.: WLAN Geräte für die Industrie im Vergleich - Teil I. Elektronik 6, 42 – 50 (March 2011)
29. Djukic, P., Mohapatra, P.: Soft-TDMAC: A Software-Based 802.11 Overlay TDMA MAC with Microsecond Synchronization. Mobile Computing, IEEE Transactions on 11(3), 478–491 (March 2012)

30. Dominguez, I., Wisniewski, L., Trsek, H.: Identification of Traffic Streams in Ethernet-based Industrial Fieldbuses. In: 15th IEEE International Conference on Emerging Technologies and Factory Automation (ETFA 2010) Bilbao, Spain (Sep 2010)
31. Dominguez, I., Wisniewski, L., Trsek, H., Jasperneite, J.: Link-Layer Retransmissions in IEEE 802.11g based Industrial Networks. In: 8th IEEE International Workshop on Factory Communication Systems (WFCS 2010). Nancy, France (May 2010)
32. Dürkop, L., Imtiaz, J., Trsek, H., Wisniewski, L., Jasperneite, J.: Using OPC-UA for the Autoconfiguration of Real-time Ethernet Systems. In: 11th International IEEE Conference on Industrial Informatics. Bochum, Germany (Jul 2013)
33. Dürkop, L., Trsek, H., Jasperneite, J., Wisniewski, L.: Towards Autoconfiguration of Industrial Automation Systems: A Case Study Using PROFINET IO. In: 17th IEEE International Conference on Emerging Technologies and Factory Automation (ETFA 2012), (best paper award). Krakau, Poland (Sep 2012)
34. Dunne, F., Pao, L., Wright, A., Jonkman, B., Kelley, N.: Combining standard feedback controllers with feedforward blade pitch control for load mitigation in wind turbines. In: Proc. 48th AIAA Aerospace Sciences Meeting, Orlando, FL, AIAA-2010-250 (2010)
35. Egea-Lopez, E., Martinez-Sala, A., Vales-Alonso, J., Garcia-Haro, J., Malgosa-Sanahuja, J.: Wireless communications deployment in industry: a review of issues, options and technologies. Computers in Industry 56(1), 29–53 (2005)
36. Elliot, E.O.: Estimates of error rates for codes on burst-noise channels. Bell Syst. Tech. J. 42, pp. 1977–1997 (Sept 1963)
37. Felser, M.: Real-time Ethernet - Industry Prospective. Proceedings of the IEEE 93(6), 1118–1129 (June 2005)
38. Freedman, D., Diaconis, P.: On the histogram as a density estimator: L_2 theory. Probability theory and Related Fields 57(4), 453–476 (1981)
39. Frotzscher, A., Wetzker, U., Bauer, M., Rentschler, M., Beyer, M., Elspass, S., Klessig, H.: Requirements and current solutions of wireless communication in industrial automation. In: Communications Workshops (ICC), 2014 IEEE International Conference on. pp. 67–72 (June 2014)
40. Gaderer, G., Loschmidt, P., Sauter, T.: Improving Fault Tolerance in High-Precision Clock Synchronization. Industrial Informatics, IEEE Transactions on 6(2), 206–215 (May 2010)
41. Gaderer, G.: Fault Tolerance Enhancements to Master/Slave Based Clock Synchronization. Ph.D. thesis, Vienna University of Technology (November 2008)
42. Gaj, P., Jasperneite, J., Felser, M.: Computer Communication within Industrial Distributed Environment - a Survey. Industrial Informatics, IEEE Transactions on PP(99), 1 (2012)
43. Gausemeier, J., Dumitrescu, R., Jasperneite, J., Kühn, A., Trsek, H.: Der Spitzencluster it's OWL auf dem Weg zu Industrie 4.0. ZWF - Zeitschrift für wirtschaftlichen Fabrikbetrieb 05/2014, 336–346 (Jun 2014)
44. Geyler, M., Caselitz, P.: Regelung von drehzahlvariablen Windenergieanlagen (Control of Variable Speed Wind Turbines). at - Automatisierungstechnik 56(12), 614 – 626 (Dec 2008)
45. Global Wind Energy Council: Global wind energy outlook 2012. online `http://www.gwec.net/wp-content/uploads/2012/11/GWEO_2012_lowRes.pdf` (November 2012), accessed: 15.08.2014
46. Golaup, A., Aghvami, H.: A multimedia traffic modeling framework for simulation-based performance evaluation studies. Computer Networks 50(12), 2071 – 2087 (2006), network Modelling and Simulation

47. Graeser, O., Trsek, H., Jasperneite, J.: Investigations on Traffic Patterns for Timing Requirements of an Industrial Real-Time System in Factory Automation. In: 2nd Junior Researcher Workshop on Real-Time Computing (JRWRTC 2008) (in conjunction with the 16th International Conference on Real-Time and Network Systems (RTNS 2008)). Rennes, France (Oct 2008)
48. Grilo, A., Macedo, M., Nunes, M.: A Scheduling Algorithm for QoS Support in IEEE802.11e Networks. IEEE Wireless Communications pp. 36–43 (Jun 2003)
49. Guan, N., Yi, W., Gu, Z., Deng, Q., Yu, G.: New Schedulability Test Conditions for Non-preemptive Scheduling on Multiprocessor Platforms. In: Proc. of the 2008 Real-Time Systems Symposium. p. 137–146. IEEE Computer Society, Washington, DC, USA (2008)
50. Gupta, R., Chow, M.Y.: Networked Control System: Overview and Research Trends. Industrial Electronics, IEEE Transactions on 57(7), 2527–2535 (July 2010)
51. Hameed, M., Trsek, H., Graeser, O., Jasperneite, J.: Performance Investigation and Optimization of IEEE802.15.4 for Industrial Wireless Sensor Networks. In: 13th IEEE International Conference on Emerging Technologies and Factory Automation (ETFA2008). Hamburg, Germany (Sep 2008)
52. Heinz, A.: Modulare Steuerungen für Windkrafträder. online `http://www.energie-und-technik.de/erneuerbare-energien/technik-know-how/windenergie/article/28782/1/Modulare_Steuerungen_fuer_Windkraftraeder/` (Aug 2010), accessed: 29.08.2014
53. Höme, S., Palis, S., Diedrich, C.: Design of communication systems for networked control system running on PROFINET. In: Factory Communication Systems (WFCS), 2014 10th IEEE Workshop on. pp. 1–8. IEEE (2014)
54. IEC: IEC 61158, Industrial communication networks - Fieldbus specifications, Part 2, Part 3-8, Part 4-8, Part 5-8 and Part 6-8 (Interbus) (2007)
55. IEC: IEC 61158, Digital data communications for measurement and control - Fieldbus for use in industrial control systems, Part 3-12, Part 4-12, Part 5-12, and Part 6-12 (Ethercat) (2008)
56. IEC: IEC 61158, Digital data communications for measurement and control - Fieldbus for use in industrial control systems, Part 5-13 and Part 6-13 (Ethernet Powerlink) (2008)
57. IEC: IEC 61158, Industrial communication networks - Fieldbus specifications, Part 2, Part 3-16, Part 4-16, Part 5-16 and Part 6-16 (Sercos I and II) (2008)
58. IEC: IEC 61158, Digital data communications for measurement and control - Fieldbus for use in industrial control systems, Part 5-15 and Part 6-15 (Modbus/TCP) (2009)
59. IEC: IEC 62591 - Industrial communication networks - Wireless communication network and communication profiles - WirelessHART (2010)
60. IEC: IEC 61158, Digital data communications for measurement and control - Fieldbus for use in industrial control systems, Part 5-10 and Part 6-10 (Profinet) (2012)
61. IEC: IEC 61158, Digital data communications for measurement and control - Fieldbus for use in industrial control systems, Part 5-2 and Part 6-2 (Ethernet/IP) (2012)
62. IEC: IEC 62264, Enterprise-control system integration, Part 1- 5 (2013)
63. IEC: IEC 61784-2:2014, Digital data communications for measurement and control - Part 2: Additional profiles for ISO/IEC 8802-3 based communication networks in real-time applications (2014)
64. IEEE: IEEE Std. 802.1q Standard for Local and Metropolitan Area Networks , Virtual bridged local area networks (2003)

65. IEEE: IEEE Std. 802.1D-2004 - IEEE Standard for Local and Metropolitan Area Networks Media Access Control (MAC) Bridges (2004)
66. IEEE: IEEE Std. 802.15.1-2005 Wireless Medium Access Control (MAC) and Physical Layer (PHY) Specifications for Wireless Personal Area Networks (WPANs) (2005)
67. IEEE: IEEE Std. 802.3-2005 Standard for Local and Metropolitan Area Networks specific requirements (2005)
68. IEEE: IEEE Std. 1588-2008 - IEEE Standard for a Precision Clock Synchronization Protocol for Networked Measurement and Control Systems (July 2008)
69. IEEE: IEEE Std. 802.15.4-2011 Wireless Medium Access Control (MAC) and Physical Layer (PHY) Specifications for Low-Rate Wireless Personal Area Networks (LR-WPANs) (2011)
70. IEEE: ANSI/IEEE Std. 802.11-2012 - Wireless LAN Medium Access Control (MAC) and Physical Layer (PHY) specifications (June 2012)
71. IEEE: IEEE Std. 802.15.4e-2012 Wireless Medium Access Control (MAC) and Physical Layer (PHY) Specifications for Low-Rate Wireless Personal Area Networks (LR-WPANs) Amendment 1: MAC sublayer (2012)
72. IEEE: IEEE Std. 802.11aa-2012 - Specific requirements Part 11: Wireless LAN Medium Access Control (MAC) and Physical Layer (PHY) Specifications Amendment 2: MAC Enhancements for Robust Audio Video Streams (May 29052012)
73. Inet: INET Framework. online `http://inet.omnetpp.org` (2014), accessed: 05.12.2014
74. Internet Engineering Task Force (IETF): RFC 2212 - Specification of Guaranteed Quality of Service (Sep 1997), `https://tools.ietf.org/html/rfc2212`, accessed: 06.10.2014
75. Jain, R.: The Art of Computer Systems Performance Analysis: Techniques for Experimental Design, Measurement, Simulation, and Modeling. John Wiley & Sons Chichester, New York, NY (April 1991)
76. Jasperneite, J., Feld, J.: PROFINET: An Integration Platform for heterogeneous Industrial Communication Systems. In: 10th IEEE International Conference on Emerging Technologies and Factory Automation (ETFA '05). pp. 815–822 (2005)
77. Jasperneite, J., Imtiaz, J., Schumacher, M., Weber, K.: A Proposal for a Generic Real-Time Ethernet System. Industrial Informatics, IEEE Transactions on 5(2), 75 –85 (May 2009)
78. Jasperneite, J., Neumann, P., Theis, M., Watson, K.: Deterministic real-time communication with switched Ethernet. In: Factory Communication Systems, 2002. 4th IEEE International Workshop on. pp. 11–18 (2002)
79. Jasperneite, J., Neumann, P.: Measurement, Analysis and Modeling of Real-Time Source Data Traffic in Factory Communication Systems. In: 3rd IEEE International Workshop on Factory Communication Systems (WFCS 2000). pp. 327–334 (September 2000)
80. Jasperneite, J., Neumann, P.: How to guarantee realtime behavior using Ethernet. In: 11th IFAC Symposium on Information Control Problems in Manufacturing (INCOM'2004). Salvador-Bahia, Brazil (Apr 2004)
81. Jeffay, K., Stanat, D., Martel, C.: On non-preemptive scheduling of period and sporadic tasks. In: Proceedings of the Twelfth Real-Time Systems Symposium. pp. 129–139 (1991)
82. Junior, R., Moraes, R., Guedes, L., Vasques, F.: GSC: A Real-Time Communication Scheme for IEEE 802.11e Industrial Systems. In: 7th IFAC International Conference on Fieldbus Systems and their Applications (FeT'07). pp. 111–118 (November 2007)

83. Just, R., Trsek, H.: Kommunikationsanforderungen an verteilte Echtzeitsysteme in der Fertigungsautomatisierung. In: Echtzeit 2009 - Software-intensive Verteilte Echtzeitsysteme. GI-Fachauschuss, Boppard (Nov 2009)
84. Kabatzke, W.: Vernetzung von Windenergieanlagen als Basis eines modernen Windparkmanagements. In: Brauer, W., Holleczek, P., Vogel-Heuser, B. (eds.) Echtzeitaspekte bei der Koordinierung Autonomer Systeme, pp. 55–64. Informatik aktuell, Springer Berlin Heidelberg (2005)
85. Kaeli, D.R.: Issues in trace-driven simulation. In: Donatiello, L., Nelson, R. (eds.) Performance Evaluation of Computer and Communication Systems, Lecture Notes in Computer Science, vol. 729, pp. 224–244. Springer Berlin Heidelberg (1993)
86. Kagermann, H., Wahlster, W., Helbig, J.: Umsetzungsempfehlungen für das Zukunftsprojekt Industrie 4.0 – Abschlussbericht des Arbeitskreises Industrie 4.0. online `http://www.forschungsunion.de/pdf/industrie_4_0_abschlussbericht.pdf` (April 2013), accessed: 01.10.2014
87. Kaliappan, P.S., König, H., Schmerl, S.: Model-driven Protocol Design Based on Component Oriented Modeling. In: Proceedings of the 12th International Conference on Formal Engineering Methods and Software Engineering. pp. 613–629. ICFEM'10, Springer-Verlag, Berlin, Heidelberg (2010)
88. Kannisto, J., Vanhatupa, T., Hannikainen, M., Hamalainen, T.D.: Software and hardware prototypes of the IEEE 1588 precision time protocol on wireless LAN. In: Proc. 14th IEEE Workshop on Local and Metropolitan Area Networks LANMAN 2005. pp. 6pp.–6 (18–18 Sept 2005)
89. Karanam, S., Trsek, H., Jasperneite, J.: Potential of the HCCA scheme defined in IEEE802.11e for QoS enabled Industrial Wireless Networks. In: 6th IEEE International Workshop on Factory Communication Systems (WFCS 2006). Torino, Italy (Jun 2006)
90. Kjellsson, J., Vallestad, A., Steigmann, R., Dzung, D.: Integration of a Wireless I/O Interface for PROFIBUS and PROFINET for Factory Automation. Industrial Electronics, IEEE Transactions on 56(10), 4279–4287 (Oct 2009)
91. König, H.: Protocol engineering. Springer, Heidelberg (2012)
92. Kopetz, H., Ademaj, A., Grillinger, P., Steinhammer, K.: The time-triggered Ethernet (TTE) design. In: Object-Oriented Real-Time Distributed Computing, 2005. ISORC 2005. Eighth IEEE International Symposium on. pp. 22–33 (May 2005)
93. Krammer, L., Seifried, S., Kastner, W.: A fault-tolerant Backbone for IEEE 802.15.4 based Networks. In: Proc. of the IEEE International Conference on Industrial Technology (ICIT 2014) (Feb 2014)
94. Kärcher, B., Heynicke, R., Scholl, G.: WSAN - Wireless Kommunikation auf der Sensor/Aktorebene auf dem Weg von der Spezifikation zur Umsetzung. In: Jahreskolloquium "Kommunikation in der Automation" (KommA 2014) (Nov 2014)
95. Krommenacker, N., Lecuire, V.: Building Industrial Communication Systems based on IEEE 802.11g wireless technology. In: 10th IEEE International Conference on Emerging Technologies and Factory Automation (ETFA 2005) (Sept 2005)
96. Kunert, K., Uhlemann, E., Jonsson, M.: Enhancing reliability in IEEE 802.11 based real-time networks through transport layer retransmissions. In: Industrial Embedded Systems (SIES), 2010 International Symposium on. pp. 146–155 (2010)
97. Kuppa, S., Dattatreya, G.R.: Modeling and Analysis of Frame Aggregation in Unsaturated WLANs with Finite Buffer Stations. In: Proc. IEEE International Conference on Communications ICC '06. vol. 3, pp. 967–972 (2006)
98. Law, A., Kelton, W.: Simulation modeling and analysis. Boston, MA [etc.]: McGraw-Hill (2007)

99. Lim, L., Malik, R., Tan, P., Apichaichalermwongse, C., Ando, K., Harada, Y.: A QoS scheduler for IEEE 802.11e WLANs. In: Consumer Communications and Networking Conference, 2004. CCNC 2004. First IEEE. pp. 199–204 (Jan 2004)
100. Lin, Y., Wong, V.W.S.: Frame Aggregation and Optimal Frame Size Adaptation for IEEE 802.11n WLANs. In: Proc. IEEE Global Telecommunications Conference GLOBECOM '06. pp. 1–6 (2006)
101. Linux Wireless: Linux wireless documentation. online `https://wireless.wiki.kernel.org/` (May 2014), accessed: 30.06.2014
102. Linville, J.W.: Tux on the Air: The State of Linux Wireless Networking. In: Proceedings of the Linux Symposium. vol. 2, pp. 39–46. Ottawa, Ontario, Canada (Jul 2008)
103. Litz, L., Gabriel, T., GroSS, M., Gabel, O.: Networked Control Systems (NCS) – State of the Art and Future. at - Automatisierungstechnik 56(1/2008), 4 – 19 (2008)
104. Lo, S.C., Lee, G., Chen, W.T.: An efficient multipolling mechanism for IEEE 802.11 wireless LANs. Computers, IEEE Transactions on 52(6), 764–778 (June 2003)
105. Loschmidt, P., Exel, R., Nagy, A., Gaderer, G.: Limits of synchronization accuracy using hardware support in IEEE 1588. In: Proc. IEEE International Symposium on Precision Clock Synchronization for Measurement, Control and Communication ISPCS 2008. pp. 12–16 (22–26 Sept 2008)
106. Lukas, G.: Fault-tolerant Industrial Wireless Mesh Network Infrastructure. Ph.D. thesis, University of Magdeburg (2012)
107. Lukas, G., Herms, A., Ivanov, S., Nett, E.: Dependable wireless mesh networks: An integrated approach. International Journal of Parallel, Emergent and Distributed Systems 24(2), 151 – 169 (April 2009)
108. Mahmood, A., Exel, R.: Servo design for improved performance in software timestamping-assisted WLAN synchronization using IEEE 1588. In: Emerging Technologies & Factory Automation (ETFA), 2013 IEEE 18th Conference on. pp. 1–8 (2013)
109. Mahmood, A., Exel, R., Sauter, T.: Delay and Jitter Characterization for Software-Based Clock Synchronization Over WLAN Using PTP. Industrial Informatics, IEEE Transactions on 10(2), 1198–1206 (May 2014)
110. Mahmood, A., Exel, R., Sauter, T.: Impact of hard-and software timestamping on clock synchronization performance over IEEE 802.11. In: Factory Communication Systems (WFCS), 2014 10th IEEE Workshop on. pp. 1–8 (2014)
111. Mahmood, A., Gaderer, G., Loschmidt, P.: Software support for clock synchronization over IEEE 802.11 wireless LAN with open source drivers. In: Precision Clock Synchronization for Measurement Control and Communication (ISPCS), 2010 International IEEE Symposium on. pp. 61–66 (Sept 2010)
112. Mahmood, A., Trsek, H., Gaderer, G., Schwalowsky, S., Kerö, N.: Towards High Accuracy in IEEE 802.11 based Clock Synchronization using PTP. In: 2011 International IEEE Symposium on Precision Clock Synchronization for Measurement, Control and Communication (ISPCS 2011). Munich, Germany (Sep 2011)
113. Malinen, J.: Hostapd Authenticator und WPA Supplicant (Jun 2010), `http://hostap.epitest.fi/`, accessed: 30.11.2014
114. Mangold, S., Choi, S., May, P., Klein, O., Hiertz, G., Stibor, L.: IEEE 802.11e Wireless LAN for Quality of Service (Invited Paper). In: European Wireless 2002. pp. 32–39 (2002)
115. Maraslis, K., Chatzimisios, P., Boucouvalas, A.: IEEE 802.11aa: Improvements on video transmission over wireless LANs. In: ICC 2012 - IEEE International Conference on Communications. pp. 115–119 (2012)

116. Matiakis, T., Hirche, S., Buss, M.: Networked Control Systems with Time-Varying Delay – Stability through Input-Output Transformation. at - Automatisierungstechnik 56(1/2008), 29 – 37 (2008)
117. Milhim, A., Chen, Y.C.: An Adaptive Polling Scheme to Improve Voice Transmission over Wireless LANs. In: Computer Systems and Applications, 2007. AICCSA '07. IEEE/ACS International Conference on. pp. 146–152 (May 2007)
118. Miller, L.: Validation of 802.11a/UWB Coexistence Simulation. Tech. rep., National Institute for Standards and Technologies (NIST) (Oktober 2003), http://www.antd.nist.gov/wctg/manet/docs/coexvalid_031017.pdf, accessed: 30.10.2014
119. Mock, M., Frings, R., Nett, E., Trikaliotis, S.: Clock synchronization for wireless local area networks. In: Proc. 12th Euromicro Conference on Real-Time Systems Euromicro RTS 2000. pp. 183–189 (19–21 June 2000)
120. Mock, M., Frings, R., Nett, E., Trikaliotis, S.: Continuous clock synchronization in wireless real-time applications. In: Reliable Distributed Systems, 2000. SRDS-2000. Proceedings The 19th IEEE Symposium on. pp. 125 –132 (2000)
121. Molina-Garcia-Pardo, J., Rodriguez, J., Juan-Llacer, L.: MIMO channel sounder based on two network analyzers. Instrumentation and Measurement, IEEE Transactions on 57(9), 2052–2058 (2008)
122. Moraes, R., Portugal, P., Vasques, F.: Simulation Analysis of the IEEE 802.11e EDCA Protocol for an Industrially-Relevant Real-Time Communication Scenario. In: Emerging Technologies and Factory Automation, 2006. ETFA '06. IEEE Conference on. pp. 202–209 (2006)
123. Moraes, R., Portugal, P., Vitturi, S., Vasques, F., Souto, P.: Real-Time Communication in IEEE 802.11 Networks: Timing Analysis and a Ring Management Scheme for the VTP-CSMA Architecture. In: Local Computer Networks, 2007. LCN 2007. 32nd IEEE Conference on. pp. 107 –116 (Oct 2007)
124. Moraes, R., Vasques, F., Portugal, P., Fonseca, J.: VTP-CSMA: A Virtual Token Passing Approach for Real-Time Communication in IEEE 802.11 Wireless Networks. IEEE Transactions on Industrial Informatics 3(3), 215 – 224 (August 2007)
125. Morell, A., Vilajosana, X., Vicario, J.L., Watteyne, T.: Label switching over IEEE802.15.4e networks. Transactions on Emerging Telecommunications Technologies 24(5), 458–475 (2013)
126. Moyne, J.R., Tilbury, D.M.: The Emergence of Industrial Control Networks for Manufacturing Control, Diagnostics, and Safety Data. Proceedings of the IEEE 95(1), 29–47 (2007)
127. Nagy, A., Ring, F., Loschmidt, P., Mahmood, A., Kastner, W.: Integration Aspects of Wireless Industrial Automation. In: Proceedings of 37th Annual Conference of the IEEE Industrial Electronics Society (IECON '11). pp. 2998–3003 (2011)
128. Naoum-Sawaya, J., Ghaddar, B., Khawam, S., Safa, H., Artail, H., Dawy, Z.: Adaptive approach for QoS support in IEEE 802.11e wireless LAN. In: Wireless And Mobile Computing, Networking And Communications, 2005. (WiMob'2005), IEEE International Conference on. vol. 2, pp. 167–173 (2005)
129. Nett, E., Herms, A., Schemmer, S.: Real-Time Mesh Networks for Industrial Applications. In: Proceedings of 17th International Federation of Automatic Control World Congress (IFAC'08). Seoul, Korea (July 2008)
130. Nett, E., Schemmer, S.: Reliable real-time communication in cooperative mobile applications. IEEE Transactions on Computers 52(2), 166–180 (2003)
131. Nett, E.: WLAN in Automation - More Than an Academic Exercise?, Lecture Notes in Computer Science, vol. 3747/2005. Springer Berlin/Heidelberg, Dependable Computing edn. (2005)

132. Neufeld, P., Meier, U., Rauchhaupt, L., Kraetzig, M.: A Unified Approach for the Assessment of Industrial Wireless Solutions. In: 16th IEEE International Conference on Emerging Technologies and Factory Automation (ETFA 2011). Toulouse, France (Sep 2011)
133. Ohm, J., Lüke, H., La, H., et al.: Signalübertragung: Grundlagen der digitalen und analogen Nachrichtenübertragungssysteme. Springer (2010)
134. Palattella, M.R., Thubert, P., Vilajosana, X., Watteyne, T., Wang, Q., Engel, T.: 6TiSCH Wireless Industrial Networks: Determinism Meets IPv6. In: Mukhopadhyay, S.C. (ed.) Internet of Things, Smart sensors, measurement and instrumentation, vol. 9, pp. 111–141. Springer International Publishing, Cham (2014)
135. Papanastasiou, S., Mittag, J., Strom, E., Hartenstein, H.: Bridging the Gap between Physical Layer Emulation and Network Simulation. In: Wireless Communications and Networking Conference (WCNC), 2010 IEEE. pp. 1–6 (April 2010)
136. Parsons, J., Demery, D., Turkmani, A.: Sounding techniques for wideband mobile radio channels: a review. In: Communications, Speech and Vision, IEEE Proceedings I. vol. 138, pp. 437–446. IET (1991)
137. Pawlikowski, K., Jeong, H.D., Lee, J.S.: On credibility of simulation studies of telecommunication networks. Communications Magazine, IEEE 40(1), 132 –139 (Jan 2002)
138. Pedreiras, P., Gai, P., Almeida, L., Buttazzo, G.: FTT-Ethernet: a flexible real-time communication protocol that supports dynamic QoS management on Ethernet-based systems. Industrial Informatics, IEEE Transactions on 1(3), 162 – 172 (Aug 2005)
139. Pei, G., Henderson, T.: Validation of OFDM error rate model in ns-3. Boeing Research Technology pp. 1–15 (2010)
140. Petersen, S., Carlsen, S.: WirelessHART Versus ISA100.11a: The Format War Hits the Factory Floor. Industrial Electronics Magazine, IEEE 5(4), 23–34 (Dec 2011)
141. PNO: Fieldbus Integration into PROFINET IO - Guideline (Version 1.0) (2006)
142. Rappaport, T., (Firme), S.B.O.: Wireless communications: principles and practice, vol. 2. Prentice Hall PTR New Jersey (1996)
143. Rauchhaupt, L., Meier, U.: Performance classes for industrial wireless application profiles and its determination. In: Emerging Technologies Factory Automation (ETFA), 2013 IEEE 18th Conference on. pp. 1–8 (Sept 2013)
144. Regnier, P., Lima, G., Barreto, L.: Evaluation of Interrupt Handling Timeliness in Real-Time Linux Operating Systems. Operating system review 42(6), 52–63 (2008)
145. Rentschler, M., Laukemann, P.: Performance Analysis of Parallel Redundant WLAN. In: 17th IEEE International Conference on Emerging Technologies and Factory Automation. Krakow, Poland (Sep 2012)
146. Romdhani, L., Ni, Q., Turletti, T.: Adaptive EDCF: enhanced service differentiation for IEEE 802.11 wireless ad-hoc networks. In: Wireless Communications and Networking, 2003. WCNC 2003. 2003 IEEE. vol. 2, pp. 1373–1378 (2003)
147. Roy, M., Jamadagni, H.: Performance analysis of MQAM-OFDM based WLAN in presence of Zigbee interference in AWGN and Rayleigh fading channel. In: Information Technology: New Generations, 2009. ITNG'09. Sixth International Conference on. pp. 1178–1183. IEEE (2009)
148. Ruscelli, A.L., Cecchetti, G., Alifano, A., Lipari, G.: Enhancement of QoS support of HCCA schedulers using EDCA function in IEEE 802.11e networks. Ad Hoc Networks 10(2), 147 – 161 (2012)
149. Santandrea, G.: A Profinet IO application implemented on wireless LAN. In: 6th IEEE International Workshop on Factory Communication Systems (WFCS 2006), (Industry Day). Torino, Italy (Jun 2006)

150. Sauter, T., Jasperneite, J., Lo Bello, L.: Towards new hybrid networks for industrial automation. In: Emerging Technologies Factory Automation, 2009. ETFA 2009. IEEE Conference on. pp. 1 –8 (Sept 2009)
151. Sauter, T., Soucek, S., Kastner, W., Dietrich, D.: The Evolution of Factory and Building Automation. Industrial Electronics Magazine, IEEE 5(3), 35–48 (Sept 2011)
152. Sauter, T.: The Three Generations of Field-Level Networks – Evolution and Compatibility Issues. IEEE Transactions on Industrial Electronics 57(11), 3585–3595 (2010)
153. Scheible, G., Dzung, D., Endresen, J., Frey, J.E.: Unplugged but connected - Design and Implementation of a Truly Wireless Real-Time Sensor/Actuator Interface. IEEE Industrial Electronics Magazine 1(2), 25 – 34 (2007)
154. Schemmer, S.: A Middleware for Cooperating Mobile Embedded Systems. Ph.D. thesis, University of Magdeburg (2004)
155. Schriegel, S., Trsek, H., Jasperneite, J.: Enhancement for a Clock Synchronization Protocol in Heterogeneous Networks. In: 2009 International IEEE Symposium on Precision Clock Synchronization for Measurement, Control and Communication (ISPCS 2009). Brescia, Italy (Oct 2009)
156. Seno, L., Vitturi, S., Zunino, C.: Analysis of Ethernet Powerlink Wireless Extensions Based on the IEEE 802.11 WLAN. Industrial Informatics, IEEE Transactions on 5(2), 86 –98 (May 2009)
157. Siemens AG: Siemens Scalance W series. online `http://w3.siemens.com/mcms/industrial-communication/en/industrial-wireless-communication/network_components/Pages/network-components-iwlan.aspx` (Oktober 2014), accessed: 04.10.2014
158. Son, J., Choi, H., Park, S.C.: An effective polling MAC scheme for IEEE 802.11e. In: Proc. IEEE International Symposium on Communications and Information Technology ISCIT 2004. vol. 1, pp. 296–301 vol.1 (2004)
159. Stanislowski, D., Vilajosana, X., Wang, Q., Watteyne, T., Pister, K.: Adaptive Synchronization in IEEE802.15.4e Networks. Industrial Informatics, IEEE Transactions on 10(1), 795–802 (Feb 2014)
160. Toscano, E., Misenti, F., Bello, L.L.: A traffic scheduler for real-time wireless communication in adaptable industrial automation systems. In: Emerging Technologies and Factory Automation (ETFA), 2010 IEEE Conference on. pp. 1–8 (2010)
161. Treytl, A., Sauter, T., Adamczyk, H., Ivanov, S., Trsek, H.: Security Concepts for Flexible Wireless Automation in Real-Time Environments. In: 14th IEEE International Conference on Emerging Techonologies and Factory Automation (ETFA 2009) (Status: accepted). Palma de Mallorca, Spain (Sep 2009)
162. Trsek, H., Jasperneite, J., Karanam, S.: A Simulation Case Study of the new IEEE 802.11e HCCA mechanism in Industrial Wireless Networks. In: Emerging Technologies and Factory Automation, 2006. ETFA '06. IEEE Conference on. pp. 921 –928. Prague (sept 2006)
163. Trsek, H., Gaderer, G.: flexWARE - Drahtlose Echtzeitkommunikation für die Fertigungsautomatisierung. In: Jahreskolloquium "Kommunikation in der Automation" (KommA 2010) (Nov 2010)
164. Trsek, H., Jasperneite, J.: An Isochronous Medium Access Control for Real-time Wireless Communications in Industrial Automation Systems - A Use Case for Wireless Clock Synchronization. In: 2011 International IEEE Symposium on Precision Clock Synchronization for Measurement, Control and Communication (ISPCS 2011). Munich, Germany (Sep 2011)
165. Trsek, H., Jasperneite, J., Lessmann, G.: A Concept for the System Integration of Wireless Sensor Networks to Industrial Automation Systems using PROFINET. In:

7th IFAC International Conference on Fieldbuses and nETworks in industrial and embedded systems. Toulouse, France (Nov 2007)
166. Trsek, H., Jasperneite, J., Lo Bello, L., Manic, M.: Industrial Electronics Handbook, chap. Wireless Local Area Networks, pp. 48–1 – 48–13. CRC Press, Boca Raton, FL, 2 edn. (Apr 2011)
167. Trsek, H., Schwalowsky, S., Czybik, B., Jasperneite, J.: Implementation of an Advanced IEEE 802.11 WLAN AP for Real-time Wireless Communications. In: 16th IEEE International Conference on Emerging Technologies and Factory Automation (ETFA 2011). Toulouse, France (Sep 2011)
168. Trsek, H., Tack, T., Givehchi, O., Jasperneite, J., Nett, E.: Towards an Isochronous Wireless Communication System for Industrial Automation. In: 18th IEEE International Conference on Emerging Technologies and Factory Automation (ETFA 2013). Cagliari, Italy (Sep 2013)
169. Trsek, H., Wisniewski, L., Toscano, E., Lo Bello, L.: A Flexible Approach for Real-time Wireless Communications in Adaptable Industrial Automation Systems. In: 16th IEEE International Conference on Emerging Technologies and Factory Automation (ETFA 2011). Toulouse, France (Sep 2011)
170. Varga, A.: Using the OMNeT++ discrete event simulation system in education. IEEE Transactions on Education 42(4), 11 pp. (Nov 1999)
171. VDI/VDE: VDI/VDE Guideline 3687: Selection of field bus systems by evaluating their performance characteristics for industrial applications (November 1999)
172. VDI/VDE: VDI/VDE Guideline 2185: Radio based communication in industrial automation (Sep 2007)
173. VDI/VDE: VDI/VDE Guideline 5600: Manufacturing Execution Systems (MES) (December 2007)
174. Vedral, A.: Digitale Analyse, Leistungsbewertung und generative Modellierung von WPAN-Verbindungen unter industriellen Ausbreitungsbedingungen. Ph.D. thesis, Brandenburgische Technische Universität Cottbus (Oktober 2007)
175. Viegas, R., Guedes, L.A., Vasques, F., Portugal, P., Moraes, R.: A new MAC scheme specifically suited for real-time industrial communication based on IEEE 802.11e. Computers & Electrical Engineering 39(6), 1684–1704 (2013)
176. Vitturi, S., Lo Bello, L.: An approach to enhance the QoS support to real-time traffic on IEEE 802.11e networks. In: 6th Intl Workshop On Real Time Networks (RTN 07) (2007)
177. Vitturi, S., Toscano, E., Lo Bello, L.: CWFC: A contention window fuzzy controller for QoS support on IEEE 802.11e EDCA. In: Emerging Technologies and Factory Automation, 2008. ETFA 2008. IEEE International Conference on. pp. 1193–1196 (2008)
178. Vitturi, S., Tramarin, F., Seno, L.: Industrial Wireless Networks: The Significance of Timeliness in Communication Systems. IEEE Industrial Electronics Magazine 7(2), 40–51 (2013)
179. Willig, A.: A MAC protocol and a scheduling approach as elements of a lower layers architecture in wireless industrial LANs. In: Factory Communication Systems, 1997. Proceedings. 1997 IEEE International Workshop on. pp. 139 –148 (Oct 1997)
180. Willig, A., Kubisch, M., Hoene, C., Wolisz, A.: Measurements of a wireless link in an industrial environment using an IEEE 802.11-compliant physical layer. IEEE Transactions on Industrial Electronics 49(6), 1265–1282 (2002)
181. Willig, A., Matheus, K., Wolisz, A.: Wireless Technology in Industrial Networks. Proceedings of the IEEE 93(6), 1130–1151 (June 2005)

182. Willig, A.: Recent and Emerging Topics in Wireless Industrial Communications: A Selection. Industrial Informatics, IEEE Transactions on 4(2), 102 –124 (May 2008)
183. Wisniewski, L., Jasperneite, J., Diedrich, C.: Effective and fast approach to schedule communication in PROFINET IRT networks. In: Industrial Electronics (ISIE), 2013 IEEE International Symposium on. pp. 1–6 (May 2013)
184. Xiao, Y.: Performance analysis of IEEE 802.11e EDCF under saturation condition. In: Communications, 2004 IEEE International Conference on. vol. 1, pp. 170–174 (2004)
185. Y.Y. Yusuf, M. Sarhadi, A.G.: Agile manufacturing: The drivers, concepts and attributes. International Journal of Production Economics 62, 33 – 43 (1999)